让-米歇尔·威尔莫特

JEAN-MICHEL WILMOTTE

建筑设计作品集

[法] 让-米歇尔 · 威尔莫特（Jean-Michel Wilmotte）/ 编著　　姜　楠 / 译

让-米歇尔 · 威尔莫特

JEAN-MICHEL WILMOTTE

建筑设计作品集

广西师范大学出版社
· 桂林 ·
images
Publishing

图书在版编目(CIP)数据

让-米歇尔·威尔莫特建筑设计作品集/(法)让-米歇尔·威尔莫特(Jean-Michel Wilmotte)编著;姜楠译.—桂林:广西师范大学出版社,2018.9
(著名建筑事务所系列)
ISBN 978-7-5598-1101-1

Ⅰ.①让… Ⅱ.①让… ②姜… Ⅲ.①建筑设计-作品集-法国-现代 Ⅳ.①TU206

中国版本图书馆CIP数据核字(2018)第202450号

出 品 人:刘广汉
责任编辑:肖 莉
助理编辑:季 慧
版式设计:吴 茜

广西师范大学出版社出版发行
(广西桂林市五里店路9号 邮政编码:541004
网址:http://www.bbtpress.com)
出版人:张艺兵
全国新华书店经销
销售热线:021-65200318 021-31260822-898
恒美印务(广州)有限公司印刷
(广州市南沙区环市大道南路334号 邮政编码:511458)
开本:635mm×965mm 1/8
印张:32 字数:80千字
2018年9月第1版 2018年9月第1次印刷
定价:268.00元

CONTENTS 目录

不可分割的联系

菲利普·乔迪迪奥

让-米歇尔·威尔莫特生于1948年，1973年从法国巴黎卡蒙多设计学院毕业后，于1975年创立了自己的公司。当他在1987年赢得了一场竞标时才受到了公众的关注——这是在位于卢浮宫拿破仑庭院内贝聿铭设计的玻璃金字塔旁的新空间，建设临时展览空间、书店和餐厅的项目。他曾谈及这些空间以及他与贝聿铭和米歇尔·马卡里的密切合作，他说："卢浮宫项目第一阶段的工作是基于一个非常周密而详细的计划来进行的。这个项目使我能与贝聿铭和马卡里的作品有一种非常有趣的对话。我想要朝着他们所做的方向前进，并牢记卢浮宫的精神内涵。我认为在这种情况下，保持一种将新建建筑完全融合到原有架构中的理念才是恰到好处的。当然，这并不是说就必须要试图以任何方式淡化我自身作品的个性。"1993年，威尔莫特回到了卢浮宫，在卢浮宫新翻修的黎塞留庭院项目中，通过与贝聿铭和马卡里合作，对面积有4800平方米的展馆进行了室内设计。与多哈的伊斯兰艺术博物馆的情况一样，他还负责了整个黎塞留庭院21 500平方米的展柜与家具的规划设计。威尔莫特不仅再一次将卢浮宫的展品最完美地呈现在人们眼前，同时也体现出了贝聿铭对于这座历史性建筑全新的室内设计。

从首尔、贝鲁特到萨尔朗

继卢浮宫项目之后，让-米歇尔·威尔莫特成为欧洲博物馆领域最为活跃和最有影响力的室内设计师之一，但是到了1993年，他也开始凭借自身实力积极参与到建筑师的事业当中，实施了众多城市的设施设计项目，例如1994年巴黎香榭丽舍大街的整修改造项目。（注：让-米歇尔·威尔莫特为香榭丽舍大街设计了休息长椅和交通灯，还设计了沿街高挑的铁质路灯等。）作为一名室内设计师，他为尚美巴黎（Chaumet，法国珠宝及奢华腕表品牌）、约翰·加利亚诺（John Galliano，法国著名服装品牌克里斯汀·迪奥首席设计师）以及卡地亚（Cartier，著名法国珠宝商及高级制表商）等知名客户进行了精品店与展厅的设计。作为一名建筑师，他完成了佳娜艺术中心（韩国首尔，1996—1998）以及两年后的首尔仁寺洞艺术中心的建筑设计工作。1999年，威尔莫特对贝鲁特国家博物馆进行了翻新改造工作，并在法国萨尔朗建造了雅克·希拉克博物馆（2002年和2006年，分两期建造），并于希拉克总统任职期间为其献礼。虽然展现政治献礼作品对于像让-米歇尔·威尔莫特这样的建筑师来说，并不是他最引人注目的典型作品，但由于他与法国这位前总统的友谊，尤其是他对萨尔朗地区建筑的理解，使他不仅能够建造出对这位总统表达敬意的建筑，同时还能使建筑的细微之处与原有的古老建筑相得益彰，但又具有强烈的现代气息。他的作品在尊重旧有建筑与必要的现代元素之间总能取得良好的平衡。威尔莫特对这些问题有着本能的把握，就像他掌握了不透明与透明或者坚固与轻盈之间的巧妙平衡一样。威尔莫特的作品最大的特点就是对于光的运用——它们总能自然而然地从看似厚重的墙壁中渗透出来。让-米歇尔·威尔莫特设计的建筑倾向于从各个角度看起来都颇有几分厚重感或具有坚实的外观，然而，光线却总是能从其建筑作品的一些隐蔽的入口过滤进来或者通过窗户照进来。尽管现代建筑经常会追求那种以缥缈、虚幻的照明来表达出一种昙花一现的感觉，但威尔莫特的建筑却像过去的石制建筑一样非常有存在感，即便它们是用混凝土和钢铁制成的。为了体现这种存在感，他加入了一种微妙的半透明或透明的、光线充足的空间元素。这种元素组合显然是矛盾的，不过也许这也恰恰能最好地诠释出威尔莫特与其他建筑师都不同的个人“风格”。他设计的建筑物外表都颇为坚固，这可能与他翻修历史性建筑方面的长期经历有关，如改造卢浮宫“原始艺术”画廊（2000年）的各种空间。

一次又一次地将过去与现在联系起来

“原始艺术（Art Premiers）”一词可以翻译为“最初的艺术（First Arts）”或“原初艺术（Primal Arts）”，其在别处的表达还有“Primitive Art”，也是指原始艺术的意思。让-米歇尔·威尔莫特与艺术品收藏界著名人物雅克·科切克合作，并在雅克·希拉克总统的领导下，将塞纳河畔的卢浮宫原始艺术厅画廊进行了翻新改造，以展出来自非洲、美洲、亚洲与大洋洲的艺术品。经过让-米歇尔·威尔莫特重新规划设计之后，来自法国国家藏品中的108件艺术品在这个面积为1400平方米的场地中进行对外展出。毫无疑问，这些艺术展示要与那些已经在卢浮宫享有盛誉的西方艺术形式旗鼓相当才行。在这种情况下，他的设计当然不仅要有不同寻常的艺术作品展示（其中一些展品甚至非常之大），而且还要表现出对于卢浮宫本身的尊崇之情。一种灵活多变的高架照明系统被人们看作是对这些艺术品的贴心设计，它使得威尔莫特成功地弥合了不同地域、不同时空文明之间的鸿沟，从这一点来说就已经不算一个小成就了。

1995年，随着卢浮宫黎塞留庭院重新开放以后，让-米歇尔·威尔莫特与贝聿铭的合作并没有结束，他们在与之前完全不同的环境中，于多哈伊斯兰艺术博物馆项目上再次携手。在这座壮观的建筑的设计中，贝聿铭在多哈滨海路一个人造岛上竖立起了他心中的“伊斯兰建筑的精髓”。这位美籍华裔建筑师将主要设计贯穿于博物馆中心的中庭空间，但展厅基本上是以一系列“黑盒子”的形式留给了威尔莫特。在这些空间中，威尔莫特设计了展柜，并精心使用了各种照明设施与材料，以将博物馆收藏的伊斯兰艺术品放置在各个角落供人们欣赏。贝聿铭的中庭空间沐浴在自然光线之中，而威尔莫特的展厅则相对较暗，这种对比使得各类展品璀璨生辉，尽显艺术之美。通过对稀有木材和宝石的运用，威尔莫特为这些展厅的方寸之间都营造出一种真正的、珍惜华贵且坚实有力的感觉，而这是非常有利于艺术品的完美展现的。

照亮名画《夜巡》

2013年完成的荷兰阿姆斯特丹国立博物馆新展览空间是让-米歇尔·威尔莫特最为重要的项目之一，但它可能并没有得到人们应有的重视。荷兰阿姆斯特丹国立博物馆建于1885年，是由荷兰建筑师皮耶特·柯伊珀斯（Pieter Cuypers，1827—1921）设计建造的新哥特式与文艺复兴风格的建筑，距离阿姆斯特丹市中心也不远。2003年，荷兰政府对国立博物馆实施了全面的重组与改造计划，以使之达到国际顶级博物馆的标准。威尔莫特设计了12 000平方米的新展览空间，西班牙克鲁兹&奥尔蒂斯建筑事务所负责包括正门厅等在内的改建重组工作。威尔莫特的主要任务是恢复柯伊珀斯原设计的内部结构与布局，并精心设计了画廊空间，以使收藏品不会触及建筑物的墙壁。建筑师对这座四层楼的建筑结构进行了改造，其中包括二楼的名画布置，如伦勃朗的《夜巡》。

他为翻新的画廊设计了不少于220个展柜。它们的透明度以及营造画廊氛围的方式，使这些展柜非常引人注目，其中的“五阶灰度”是由飞利浦照明开发的独特LED照明系统营造出来的，该照明系统是由威尔莫特专门为博物馆而设计的，名为“Lightracks”。那些荷兰国立博物馆的17世纪杰作典藏常常沐浴在这种微妙的光影中，有了这种层次丰富的阴影，它们终于在威尔莫特重新设计的画廊中找到了合适的安放之所。让–米歇尔·威尔莫特绝对是改造此类空间的大师之一，他总是能在这种改造工作中使建筑设计、室内设计和照明设计等在这类艺术品展览空间中相得益彰。以上也只能部分地说明让–米歇尔·威尔莫特在博物馆室内设计、展位设计和场景物件设计、城市规划或建筑设计等各种领域的广泛涉猎而已。事实上，有一点似乎很清楚——那就是威尔莫特的展位设计与场景物件设计和他的建筑设计风格之间有着不可分割的联系。毫无疑问，同样的精神内涵既能体现在其建筑作品的坚固外墙之上，也能够体现在其场景物件作品的可移动世界之中。和他的建筑一样，威尔莫特的场景物件几乎总是展现出坚实感甚至是沉重感，然而它们也会被赋予一种舒适感，并且以最为积极与最为现代的方式，赋予建筑一种坚不可摧之感。这就形成了一种“威尔莫特风格”——在过去与现在之间、轻盈与厚重之间、不透明与透明之间总是存在着大量的联系。

牢固的连接

当进行建筑设计时，他总是尊重项目所处的环境，同时也要保留基本的现代性元素。光线与透明、半透明、不透明表面的巧妙并置是他设计方法的要素之一。另一方面，他在与客户和决策者们打交道时也颇有亲和力。让–米歇尔·威尔莫特似乎认识建筑界、设计界和艺术界中的每个人，他在法国政界、商界的精英圈子里面也都有着类似的联系。无论是建筑设计还是展位设计，他都将自己的感受融入到现代建筑或场景物件的典型特征之中。尽管事务所已经拥有200多名员工，但他仍然是其建筑事务所的重要推动力量，为他们赢得各种项目，寻找适当的解决方案来解决一些非常困难的建筑设计问题，如将那些古老的欧洲石制建筑与现代风格相互融合在一起。

尽管在博物馆方面的工作一直都是让–米歇尔·威尔莫特的事业亮点之一，但近年来他也主要负责了一系列其他大型项目。例如安联里维耶拉球场项目（法国尼斯，2013年），他在指定的“绿色生态区”内设计建造了这个建筑，使用了大量木材，使得球场与周边环境保持了一致性。事实上，它号称是有史以来建造的最大规模的木材与金属空间框架结构。球场的明亮通透是指导其设计的关键理念。球场上方有一个非常大的圆角的矩形开口，可以在尼斯的恶劣天气或强烈日照出现时为观众坐席提供遮蔽。

通过这种方式，球场的露天部分与遮盖部分结合在一起，下方建筑立面的半透明表皮展现出球场朦胧的骨架结构，人们可以在其形成的光影之下观看比赛。尽管球场位于地震活动区，但他还是设计了一个46米高的顶部悬臂结构。白天，球场里充满了漫射的自然光线，而在夜间，它又从内部散发着光芒。球场顶部EFTE（乙烯–四氟乙烯共聚物）膜构建出的起伏形状，给人一种流动之感，看起来就像一大块保护性遮蔽物上扬起来在迎接着观众。威尔莫特才华的一个重要方面就在这个项目中显露出来，使他的客户们感到非常安心。球场与威尔莫特其他新建的九个建筑作品相比采用了更为轻盈的结构，但它还是保留了其他作品风格中那种条理分明的现代感。球场内部舒适宜人、采光良好、视野开阔、通道很多。这座体育场旨在为各种活动提供良好的配置，包括体育赛事、音乐会和其他类型的大型公共活动等。

从塞纳河到巴比松

威尔莫特设计的俄罗斯东正教精神文化中心（法国巴黎，2016年）更容易被普通大众所了解，它坐落在塞纳河河岸附近，距离以俄国沙皇亚历山大三世命名的、具有象征意义的亚历山大三世大桥不远。该建筑群包括一个教区大厅、一个宗教中心、一所学校以及最重要的圣三一大教堂——这座大教堂以五个东正教特有的镀金圆顶为标志。该中心采用了与邻近的夏乐宫（位于河对岸）相同的石材，威尔莫特将其设计融入相当复杂的建筑环境和历史背景之中，同时也展现出了一种坚实的现代感。2012年，另一位建筑师拒绝了这项计划，之后在巴黎市长贝特朗·德拉诺埃打来电话请求帮助时，让–米歇尔·威尔莫特展示了他驾驭复杂政治环境的能力，从而以饱满的热情出色地完成了这一建筑挑战。

让–米歇尔·威尔莫特最近的其他作品已经从颇具趣味性的私人住宅设计转移到了办公大楼方面。其建筑项目的范围清楚地表明了他的能力范围——从小到大，从翻新改造到新建建筑无所不有。他为法国艺术家让–皮埃尔·雷诺（Jean Pierre Raynaud，法国巴比松，2009年）设计了住所。

1939年出生的让–皮埃尔·雷诺过去一直在以自己的方式来摆脱巴黎的喧嚣。例如，在距离拉加雷讷科伦布市中心仅10千米的地方，雷诺居住在自己设计的房屋中（名为马斯塔巴Le Mastaba，与建筑师让–德迪乌Jean Dedieu于1986年合作建造）。在拉加雷讷科伦布，雷诺创作并展出自己各种各样的艺术作品（从镀金叶的巨型花盆到通过他的改造而成为艺术作品的外国旗帜等）。后来，雷诺在让–米歇尔·威尔莫特的帮助下于巴比松建造了一个用于工作的新住所。巴比松距离巴黎58千米，被誉为巴比松画派的所在地，以泰奥多尔·卢梭和让–弗朗索瓦·米勒等19世纪的重要画家而闻名。建筑师拆除了一栋名为霍顿斯葡萄园的老房子的后期增建部分——这所20世纪60年代的房屋延伸部分，并增加了一个大型的玻璃内部空间，建造了475平方米的阳台。作为1939年世界博览会的一部分，这座拥有100年历史的小型北欧式亭屋，应艺术家的要求被全部涂成了黑色，这也是该项目整体构成的一部分。“玻璃墙”也将房子向树木繁茂的环境与艺术家自己的占地4000平方米的花园敞开，雷诺和威尔莫特与景观设计师奈维欧・卢耶合作营造出了他所钟爱的那种艺术、建筑与自然的和谐统一。事实上，让-皮埃尔・雷诺长期从事建筑与艺术之间的工作，他在1969年就在巴黎附近的拉塞勒-圣克卢建造了自己的房子，该房子直到24年后才被拆除。雷诺于1993年用拆除下来的瓦砾碎石在波尔多CAPC当代艺术美术馆组织了一次著名的展览。在这种特殊情况下，通过与让-米歇尔・威尔莫特的密切合作，这位艺术家工作室所在的建筑物也被视为他艺术作品的一部分。

声望日隆

威尔莫特积极探索高楼大厦领域的设计与施工。例如，蒙特卡洛观景塔大厦（摩纳哥，2012年）是一座三栋相连的20层高层公寓建筑，以及紫罗兰大厦（摩纳哥，2018年）则是一座位于摩纳哥东部圣罗马地区的130米高（22层）、拥有73间豪华公寓的住宅大楼。与开发商和客户以及官员们的广泛联系使他能够在许多其他建筑师不被允许涉足的地方进行设计与建造工作。他在摩纳哥进行建筑设计时就是这种情况，最近在美国亦是如此。他的哈伍德村项目（珀夏尔大厦，美国德克萨斯州达拉斯，2017年）对于这位以巴黎为大本营的建筑师来说，标志着新的一步——美国市场对他的接受程度，这个市场对于外国设计师或法国设计师来说是相当封闭的。威尔莫特事务所的涉猎范围从最先进的照明设计到具有混合用途的大厦应有尽有，但建筑师本人还是重点为大型组织机构与公司建造办公大楼。本书列举的两个案例就反映了建筑师近期在这方面的工作。法国近年来最受关注的官方项目之一就是在巴黎第15区——巴拉尔地区的法国国防部新大楼。让-米歇尔・威尔莫特的任务是为该建设地点象限角服务性园区建造各类建筑。该项目建筑面积接近100 000平方米，占地面积2.8公顷，可以轻松地呈现出一种厚重感，但威尔莫特却成功地使用了一种能让人回忆起传统巴黎联排别墅的方法，使在这座城市中的这些服务性建筑给人一种如家一般的亲切感，而办公楼的设计则给人以利落之感。

屋顶与阳台的绿色植物以及充足的自然光线使办公室变得更为舒适，而办公室还能调整大小尺寸，这可以为未来的需求提供最大程度的灵活性。在巴黎的韦利济-维拉库布莱区，威尔莫特在2015年为法国主要土木工程公司埃法日集团建造了总面积达24 000平方米的总部。这座综合设施的严肃风格以及功能方面确实颇具威尔莫特作品的特点。尽管近年来他已经尝试过一些更为活泼的形式，但他的总体风格还是倾向于以一种现代建筑中有时会被遗忘的方式来排布那些相对素雅的材料以表现建筑的坚实感。与他设计的其他地方一样，此处的案例中，一座花园像个天然的植物大碗一样排布在建筑物的周围，而建筑则采用坚硬的材料如混凝土以及黑色的金属门与窗框来进行调和。威尔莫特建筑物的材料坚实性（尤其是他设计的办公楼），总是能够通过自然光线或绿色植物来缓和。这些元素使办公楼成为更加令人愉悦的存在，它们总是会存在于现代而节能的建筑物内部及周围。

威尔莫特还找到了与法拉利这样的客户携手前行的方法，他为其建造了法拉利运动跑车管理中心（意大利马拉内罗，2015年）。法拉利运动跑车管理中心致力于设计和组装法拉利一级方程式赛车，是一座具有强烈辨识度的标志性建筑。法拉利标志性的红色、黑色和金色以及建筑表皮的条带环绕形状，都有助于营造出建筑物的力量、特色，给人一种跑车般的运动之感。建筑将法拉利品牌特征中的速度、性能和加速度等转化为一种形象的空间诠释，使得这座建筑转变成一个玻璃包裹的“容器”，如同该公司出品的鲜红色赛车一样。

威尔莫特设计的法国葡萄酒厂则是其涉猎另一个领域并赢得赞誉的项目。他设计的百德诗歌酒庄（波亚克产区，2015年）在本书中有介绍，但他也曾为罗兰百悦香槟（法国马恩河畔图尔，2015年）工作过，在那里他被要求为杰出的品牌罗兰盛世香槟设计相关设施。威尔莫特将他对建筑空间的知识、照明与室内设计融合在了一起（比如罗兰百悦250平方米的品酒室）。优质葡萄酒具有将它们的历史与现在及未来联系起来的能力，在这方面，恐怕罗兰百悦不可能找到比让-米歇尔·威尔莫特更好的建筑师和室内设计师了。2015年，威尔莫特全面翻新了圣昂托南迪瓦的酒庄，他并没有涉足新建筑，而是重新定义了现有的空间和起源于11世纪的农场建筑。同样，威尔莫特在过去相关工作中的卓越表现在这里也是显而易见的。

让-米歇尔·威尔莫特负责了巴黎十三区F站初创企业园区（巴黎，2017年）的翻新改造工作。在这个案例中，他的客户是法国自由电信电话网络的所有者泽维尔·尼尔，他也是《世界报》的部分持有人。尼尔在法国企业界享有盛名，他主动帮助那些有前途的，吸引了新闻界大量关注的创业公司。这个项目设置在由著名结构工程师尤金·弗雷西内于1927年建造的有盖大厅内。自2012年以来，它被列为历史古迹，长度为304米，宽度为61至72米。作为附近巴黎奥斯特里茨火车站的换乘点，这座大厅被威尔莫特顺便利用起来，改造后反映了年轻的数码创业者们的生活方式与商业活动。F站初创企业园区现在拥有了共享的多用途空间以及围绕在各种服务设施周围的“村落”。有盖大厅的东南端设有一间与众不同的餐厅，24小时开放。这里保留了旧的装卸平台和钢轨底座，甚至还有几节餐车，使人能够回想起过去的车站。在这间人来人往、令人惊喜的餐厅可以通过花园的阳台眺望城市的风景。威尔莫特采用了预应力混凝土、钢铁、木材和玻璃等建造了这座33 000平方米的建筑物。让-米歇尔·威尔莫特表示：“我们的数码化孵化器项目是一个真正的催化剂，它将独特、创新、充满活力的空间中的两个主要创造性源泉集中在了一起——将一位19世纪工程师的大胆才华与原生肆意的新生代想象力融合起来。”最重要的是，让-米歇尔·威尔莫特展现出了一种真正的能力——将过去的建筑风格，无论是古代的还是近代的都融入现在的建筑功能当中。诚然，拥有开放式设计的巴黎13区F站孵化器可能比威尔莫特已经处理过的一些石制旧建筑更容易应付，但是他这种将过去与现在相结合的能力在这里再次得到了证明。不同于许多建筑师以建筑外表或特别的建筑形式来明确宣示自己的建筑风格与特权，威尔莫特显然更具多方面的思考，包括与外部无缝融合的室内设计理念等。在他的大部分项目中，建筑的物件、灯具、颜色、表面和材料都与建筑设计本身完全整合成了一个整体。他驾驭室内空间的能力和对21世纪风格进行的反思都意味着他是以一种建筑设计的方式进行思考的，即使只考虑现有建筑墙壁内的空间亦是如此。

巴黎的五星级酒店

威尔莫特的巴黎文华东方酒店与水疗中心项目（法国巴黎，2011年）是将原来的办公大楼改造成一家五星级酒店。威尔莫特不仅成功完成了这项任务，而且该项目完工后还获得了法国绿色建筑标准认证（HQE）。与威尔莫特的许多项目相反，这里原来的办公大楼并不起眼，但威尔莫特在此营造出了豪华的空间——之前这里只有办公桌和走廊而已。本书付印之时，另一个大型酒店项目正在进行——让-米歇尔·威尔莫特被委托对巴黎左岸唯一的“宫殿式”酒店——鲁特西亚酒店进行全面翻新与重组工作。这个项目不是一次翻新改造那么简单——尽管该酒店在拉斯帕尔大道和塞夫勒街拐角处的外墙被列为历史古迹，但其原有的内墙却几乎荡然无存。威尔莫特在低于基准标高以下进行了挖掘工作，以构造出一个温泉和游泳池，打开了原来的一个内部庭院使之成为一个冬季花园，用其标志性的照明设施与场景物件重新布置了房间。威尔莫特在鲁特西亚酒店项目中不仅展示出了他的全方位才能，还号召各类专家根据设计需要恢复建筑物内的历史性装饰。那些在改造之前就了解鲁特西亚酒店的人不仅会发现建筑的外立面经过了翻新，还会发现有足够数量的内部元素也都焕然一新，而那些新的参观者们则会想象着这座建筑过去的辉煌并更加喜爱和享受这座酒店的现代感。同样，建筑师将过去与现在混合在一起，以便在一座五星级酒店的环境里赋予“豪华”这个词一种新的内涵。

在莫斯科的河岸上

尽管各个方面资金都显不足，让-米歇尔·威尔莫特还是接手了一些城市规划设计方案，比如他的让-饶勒斯大街项目（法国尼姆，2013年）——这座法国南部城市最大的街道。他设计了亭子与绿廊装点着宽阔的让-饶勒斯中央步行街，利用古老茂密的荨麻树构成了各种图案。两侧的外墙由采用LED背光照明的冲孔薄铁板组成，通过提供普罗旺斯花边样式的精致照明使这条大街变得生气勃勃。“我们想将尼姆喷泉花园的活力引到大街上，让人们在那里也可以感觉惬意，并更乐意去那里悠闲地漫步。”建筑师说。当然，尼姆的历史十分悠久，可以追溯到罗马时代，因此这里的改造工作充满了政治与审美上的纷争。威尔莫特再次将这些问题掌控于股掌之间，就像他在设计方面所做的那样，将这里过去与现在之间的联系几乎天衣无缝地交织在了一起。

更为雄心勃勃的是，他的大莫斯科设计方案（俄罗斯联邦，2012年）涉及到俄罗斯最大的城市——首都莫斯科。2012年9月5日，一个评审委员会授予由城市建筑师安东尼·格伦巴赫、让-米歇尔·威尔莫特和谢尔盖·特卡琴科领导的法俄团队进行莫斯科扩城规划的设计任务。应对莫斯科城市改造这种规模的项目对让-米歇尔·威尔莫特来说是一次实质性的飞跃。他与合作伙伴一道，展现出了他对这座城市的想象力，就像他对自己设计的那些建筑的想法一样，方案具有威尔莫特风格中不可磨灭的逻辑性和对建筑透明度与现代元素的良好运用。法俄团队规划方案的指导原则之一是围绕莫斯科河发展大都市区的公共机构与休闲活动设施建设。就像巴黎的塞纳河和伦敦的泰晤士河一样，莫斯科河也将因此成为这座城市特征的载体。但实际情况并非如此乐观，不过即便如此，该方案看起来仍像是由莫斯科及其居民们提出的一个“自然而然”的提案一样，很接地气。由于对生活质量与原有城市遗迹价值的持续关注，以及这里对于行人而言存在着出行难的困惑，莫斯科河河岸的景观开发将与公共交通的重新部署（在这里体现为有轨电车）相伴而生，以限制日益增长的交通流量。对于这个可能涉及155 000公顷土地的设计理念而言，威尔莫特和格伦巴赫认为，这样一个雄心勃勃的计划的关键是要抓住现有的发展路线并不断对其进行优化，而不是总是试图扭转现有的模式。在这方面，大莫斯科计划反映了典型的威尔莫特对现存问题的看法，对过去历史的尊重和对未来的开放性思维。

让-米歇尔·威尔莫特肯定不是包豪斯学派和格罗皮乌斯所说的“白板理论”（建立在拆毁得空无一物的城市之上）的支持者；相反，他似乎立即就确定了原有系统的优势和潜力，并将这些现实存在整合到与当前城市里具有同样敏感性的现代化建设中。对于威尔莫特来说，过去与现在之间并没有根本性的冲突。他从来没有沉湎于过去，也没有真正努力去还原过去，而是保留旧时那些有用而迷人的东西以增强和巩固当下的项目。在这方面，他可能是现代建筑设计与室内设计界当中独一无二的人物了。从陈列设计开始，威尔莫特将这种设计迅速扩展到大量的城市设计当中，比如他在巴黎香榭丽舍大街的城市设施设计与灯光设计方面的工作（1994年）就是如此。威尔莫特建筑事务所近年来已经成为世界100强建筑事务所之一，完成了从安联里维耶拉球场到私人住宅等各种重大项目。酒店、企业总部大楼、葡萄酒庄和博物馆等也都在他的项目清单之中。事实上，威尔莫特拥有室内设计、陈列设计和照明设计的基础，这使他在需要时也具有很大程度的灵活性来承担建筑内部的设计工作。在让-米歇尔·威尔莫特的作品中，“坚实性”这个词是最具积极含义的，但其实这也是他对自然与光线的精妙理解之所在。他从一开始就表现出了通过自然环境与人造光使不透明的建筑体鲜活起来的能力；例如，他对那些旧石制建筑的各种翻新改造案例就是如此。值得一提，威尔莫特在法国国内也同样佳作不断：他为一座11世纪的农舍增添了活力，使1927年的铁路设施成为最先进的技术创业公司孵化器，为享有盛名的法拉利车队建造了一座大楼。威尔莫特的成功来源于他对建筑设计与室内设计的精妙理解，来源于他调和过去与现在的才能，以及他那和蔼可亲而广受欢迎的个人魅力。

NEW BUILDINGS
建成项目

李氏宅邸

李氏宅邸位于韩国首都首尔平昌洞的一个居民区，该建筑位于山脚下，由钢筋混凝土建造。建设地点是在树木繁茂的山腰斜坡上，因此可以远离城市的喧嚣。威尔莫特与景观设计师弗朗索瓦·奈维欧和伯纳德·卢耶合作设计的一系列阳台和花园让建筑的混凝土构造看起来像是从郁郁葱葱的山坡中生长出来的一样。上层阳台上的梯级游泳池沿着建筑轮廓向下延伸。

李氏宅邸坐落于占地1914平方米的场地上，建筑面积达348平方米。花园里有典型的韩式古陈设。不断变化的光线照在建筑上，精心挑选的松树为房屋的墙壁增添了勃勃的生机。这座房子是个基于方块的组合体，由上面带有黑色铝制窗格的、界线明确的简洁建筑体组合而成。玻璃和混凝土表面相互交替，融合了玻璃的空洞感和混凝土的坚实感。

这所房子为业主及其家人提供了家庭生活中的温馨亲切之感。同时，该方案避免了任何复杂混乱的情况出现。起居室、餐厅、厨房和办公室位于建筑下层，有三间卧室、一间书房、一间浴室在上层。住宅空间简单而宽敞，凹进的接缝凸显了各个建筑体与各种材料在涂装混凝土区域的连接。就像建筑师所说的那样，“这是一页书写日常生活的白纸”。这座房子结合了威尔莫特的设计才能，他使这座建筑基于几何形式建造，外表面界限分明，同时又利用光和自然环境使建筑的棱角直线变得更为柔和。即使是在花园阳台中放置了古式的祖先雕像，也与这种极具包容性的现代建筑相得益彰。

项目地点 / 韩国首尔
完工时间 / 2005年
建筑面积 / 348平方米

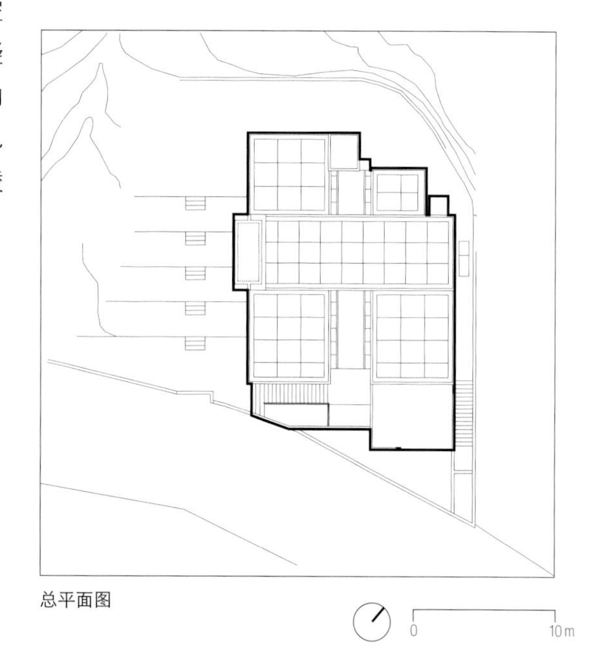

总平面图

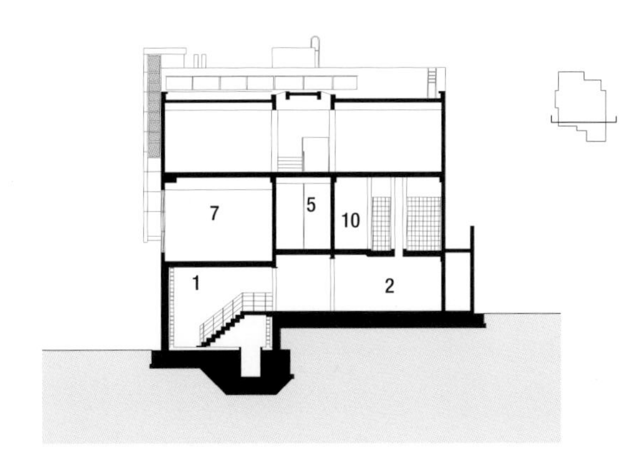

横向剖面图

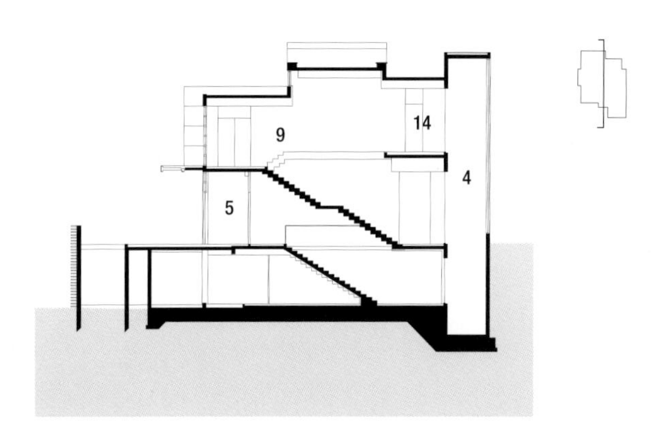

纵向剖面图

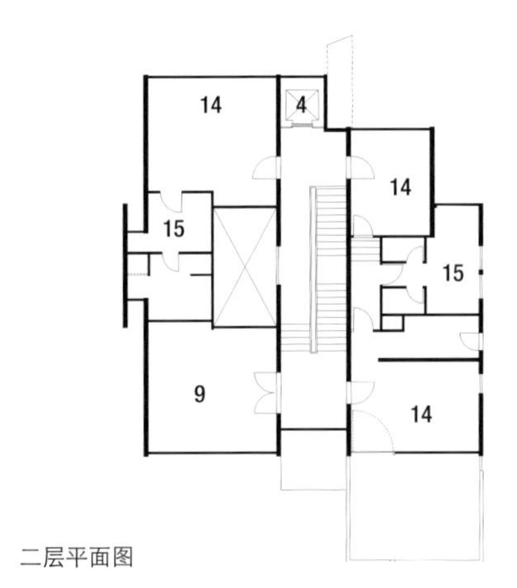

二层平面图

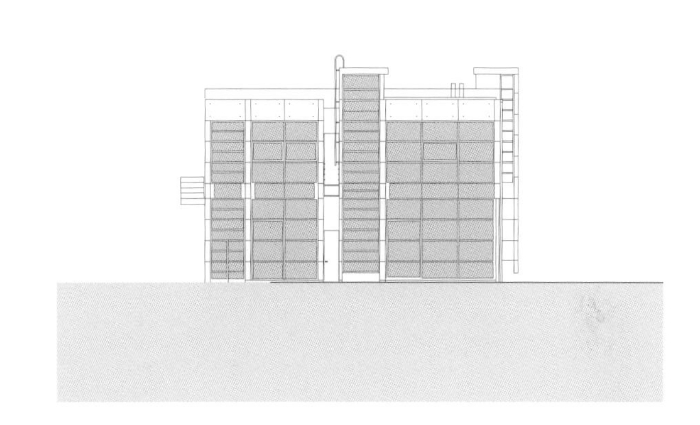
西北侧立面图

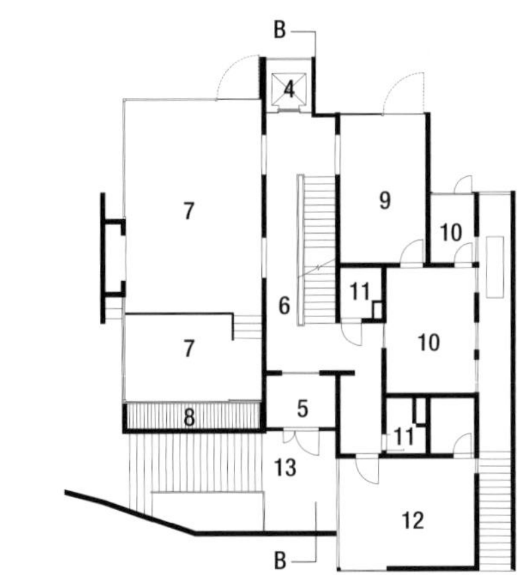

一层平面图

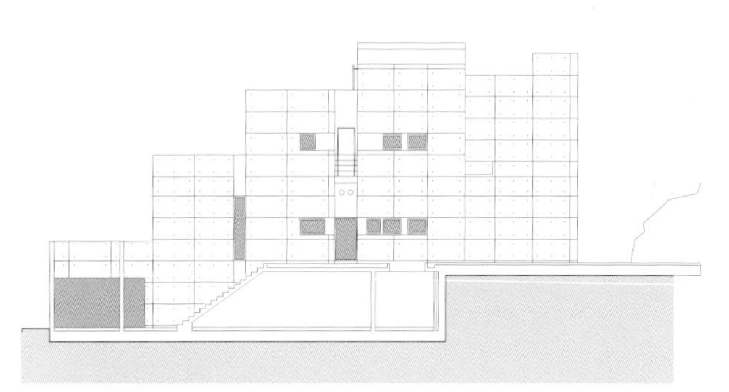
东南侧立面图

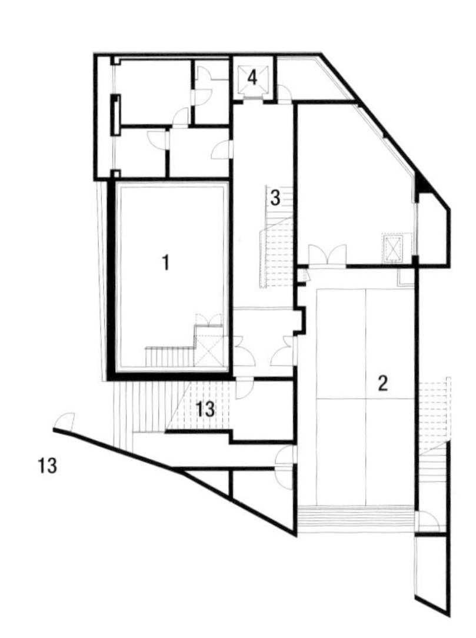

地下室平面图

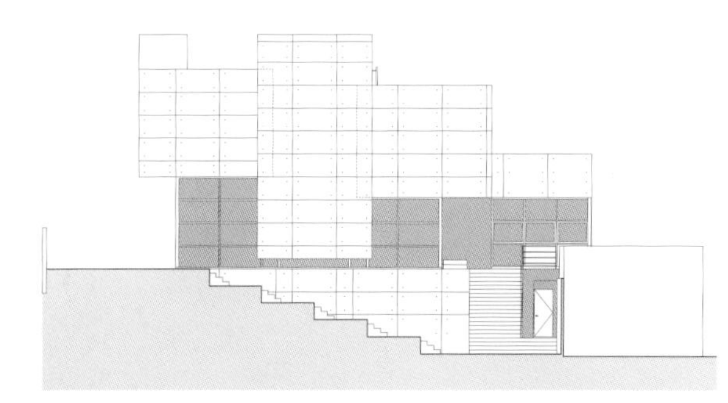
西南侧立面图

1 储藏间
2 停车位
3 室内楼梯
4 电梯
5 入口
6 门厅
7 客厅
8 门廊
9 办公室
10 厨房
11 厕所
12 电视房
13 通道
14 卧室
15 浴室

板桥高档住宅

这座豪华公寓大楼位于韩国首都首尔南郊的京畿道，占地面积约合9690平方米，由七7栋建筑组成，每栋有4层，总共有36间公寓，并建有地下车库。这座钢筋混凝土建筑的地面上采用了白色卡拉拉大理石和传统的韩国青砖。

威尔莫特再次与景观设计机构奈维欧–卢耶携手，他们提出在面积为1公顷的土地上将水体与绿色植物精心组合在一起。该项目与佳娜艺术中心的客户相同。这位法国建筑师在韩国倍受尊敬，在施工前，由威尔莫特对项目做出了介绍之后，像这样的公寓马上就被卖出了17套，买主都是体育名将、演员、舞蹈家、视觉艺术家和电影制片人等。

这些高档公寓平均面积大约为200平方米。公寓设有门卫、健身中心和各种共享服务设施。7栋建筑散落在各处，有些坐落在小区主轴上，而有的则与之垂直布置。4个最小的建筑是成对的，每层有一间公寓。第一对建筑附近的入口在小区轴线上形成了一个巨大的拱门。第二对建筑垂直于轴线一侧排列，并设有一条共享的走廊。3座最大的建筑物首尾相接，每层有两个公寓，共用一个楼梯平台。在中间小路的两侧是一片穿过小区轴线的竹林。从车库到公寓，一路上都是绿色植物与景观，令人赏心悦目。玻璃电梯始于从水池溢出而产生的一道水墙，小区的一些建筑物也被设置在这里。在每个楼梯平台，各个公寓都可以通过耸立在住宅上方的人行天桥（兼阳台）进入各自的景观庭院之中。每间公寓的设计都像一个花园别墅，它们其中的大部分都是开放式的。

项目地点 / 韩国首尔
完工时间 / 2005年
建筑面积 / 9690平方米

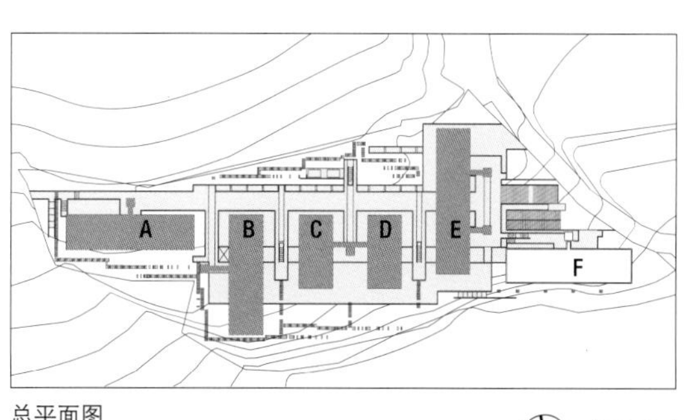

总平面图

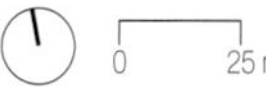

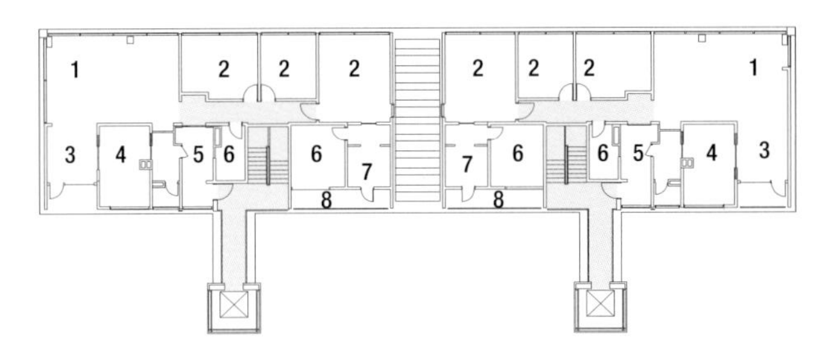

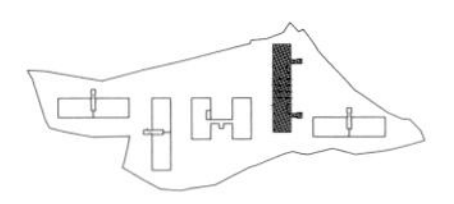

D座：四层平面图

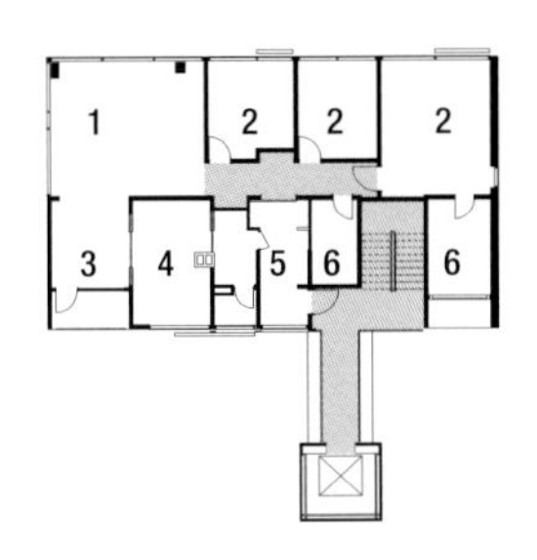

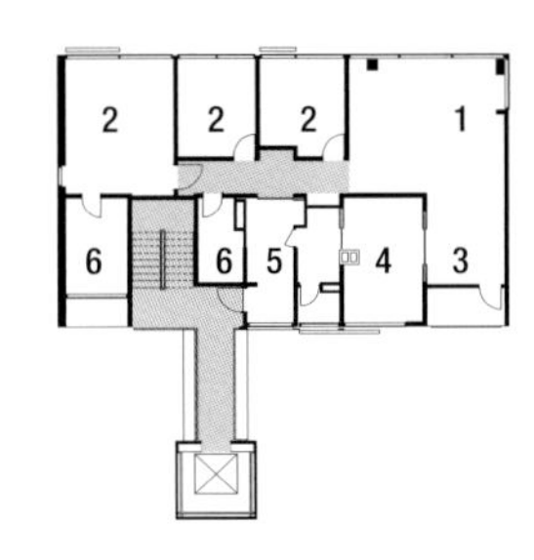

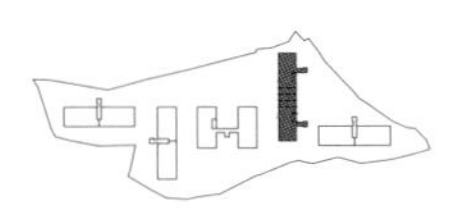

D座：三层平面图

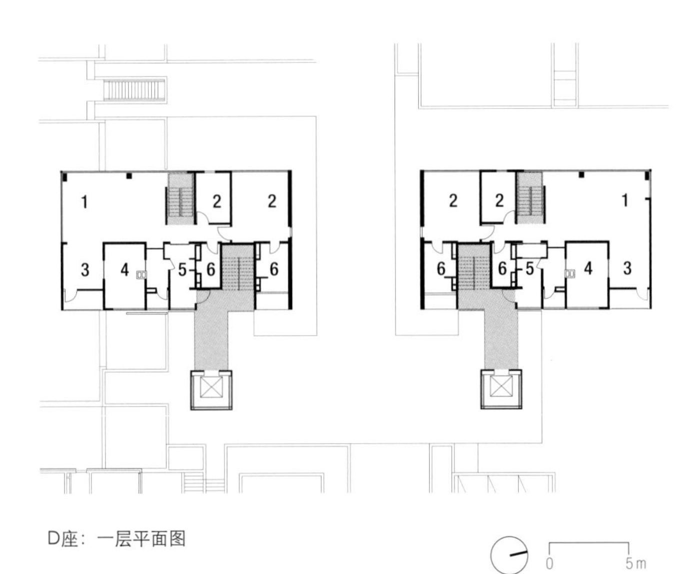

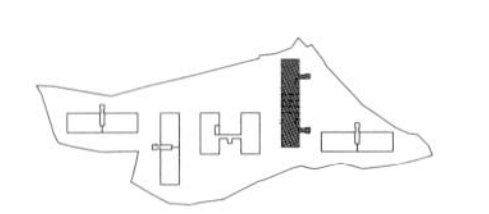

1 客厅
2 卧室
3 饭厅
4 厨房
5 入口
6 浴室
7 更衣室
8 阳台
9 健身中心
10 停车场
11 服务设施

D座：一层平面图

0 5m

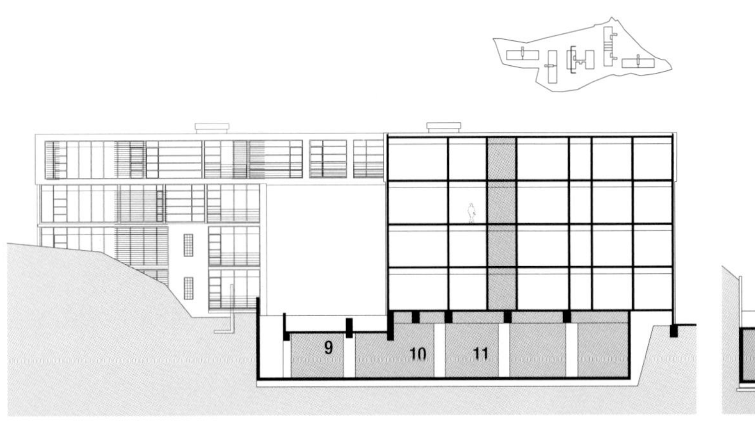

C座：横向剖面图

A座：横向剖面图

0 10m

信号塔大厦

项目地点 / 法国巴黎拉德芳斯
完工时间 / 2008年
建筑面积 / 115 000平方米

信号塔大厦是一座284米高的，位于拉德芳斯商业区的混合用途建筑，坐落在巴黎香榭丽舍大街的中轴线上。总建筑面积为115 000平方米，大厦包括办公室（31层）、公寓（10层，60间公寓）、一个旅馆（24层，230个房间和40个套房）、住宅（12层，150户住宅单元）、商店和公共设施等（5层）。

大厦的基础部分约80%都可以通达公共场所，该设计试图释放地面层的最大面积以利行人从中通过。三个构成大厦的独立元素被置于共同的底座之上，它们就像三根玻璃立柱，交错螺旋上升，时而分开，时而交会，最后在不太远的距离上又分开相对。整个大厦的旋绕设计在多个方向上强调了“信号”的理念，而从不同角度看，各个建筑体都像是从另外的笔直大厦中以一定角度分身出来一般。这是一个牢固而精巧的开放式架构，它呈现给人们一个雄心勃勃的解决方案，满足了建筑的现代性、功能性和生态性等各种需求。

2008年，信号塔大厦被列为世界上首个使用自然通风的高层建筑，同时它自身还能产生三分之一的能源供给。大厦的能源自给主要有三个来源：地热、风能和光伏板。雨水回收也是该设计方案的一个组成部分，而且方案基于建筑的三角形基座对日照进行了精准计算，可以为大厦提供最大量的自然光。这座占据了拉德芳斯三分之二场地的建筑将成为该地区最大的绿色空间，而这或许对于世界上任何一个大型城市商业区来说都意义深远。信号塔大厦作为一个资金充分，与现有公共交通体系完美结合的项目，体现了巴黎包容万象的精神，为人们提供了开放交流的理想之所。

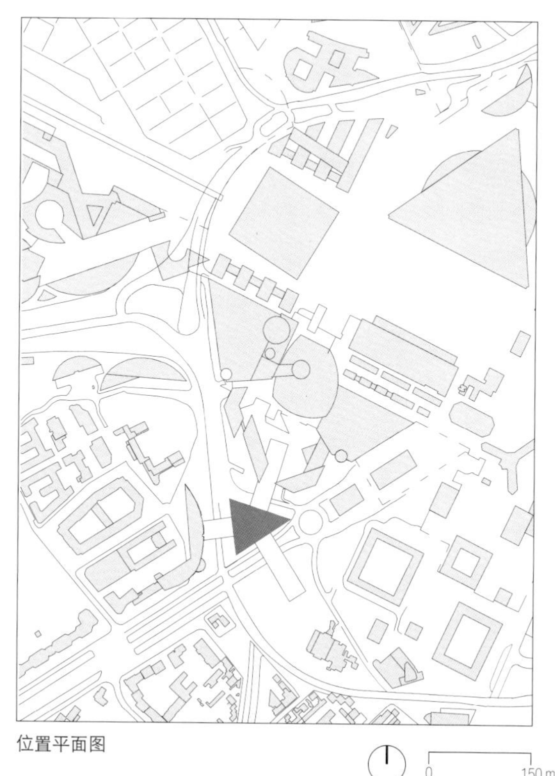

位置平面图

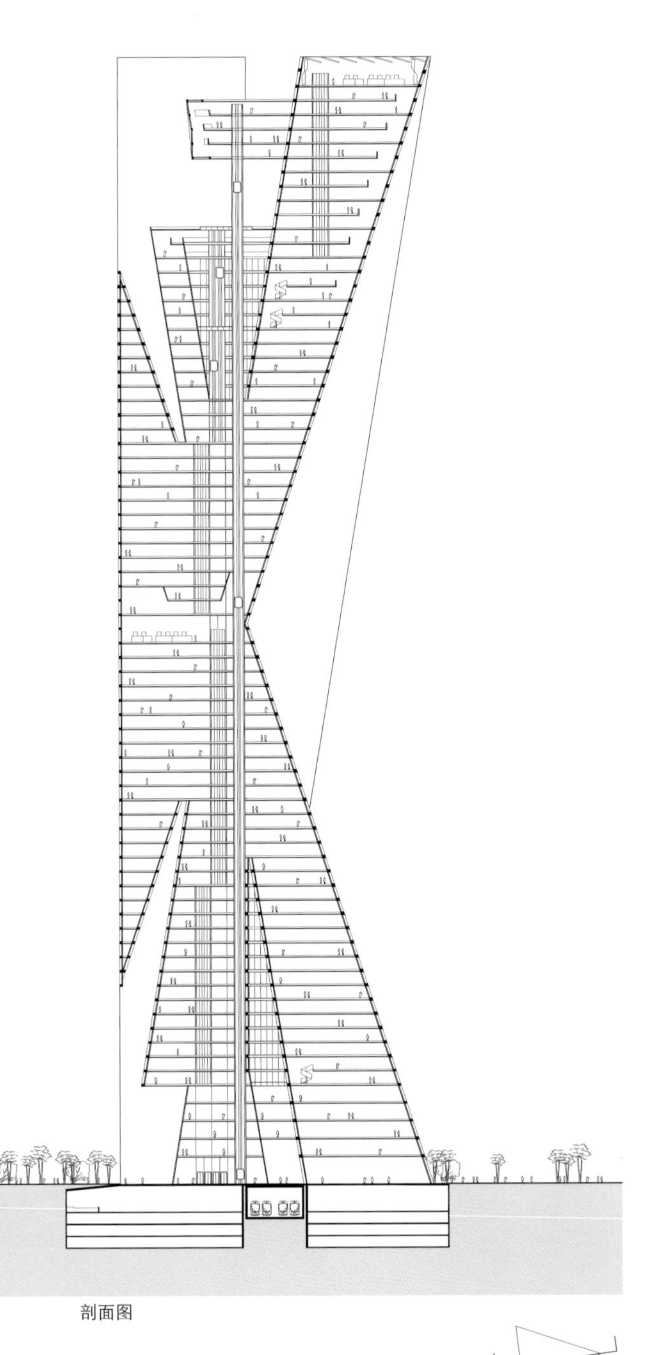
剖面图

公寓层平面图（74层）

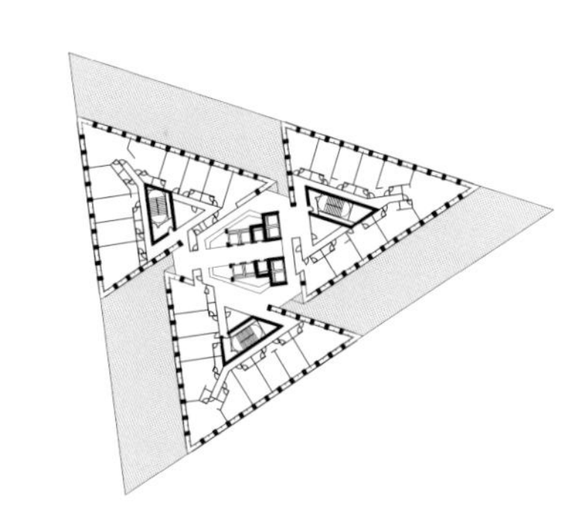
酒店层平面图（57层）

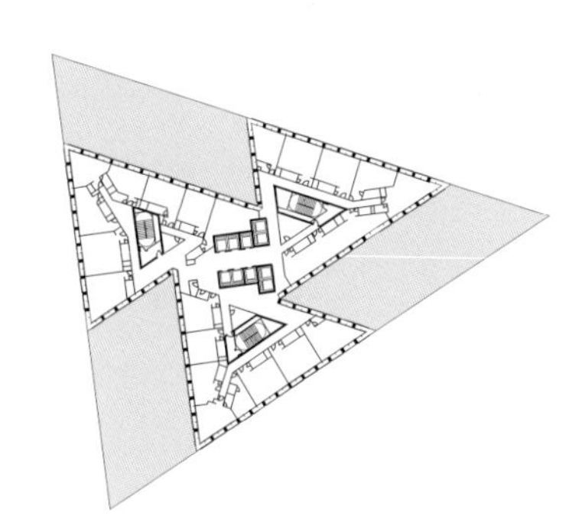
商旅观光房平面图（51层）

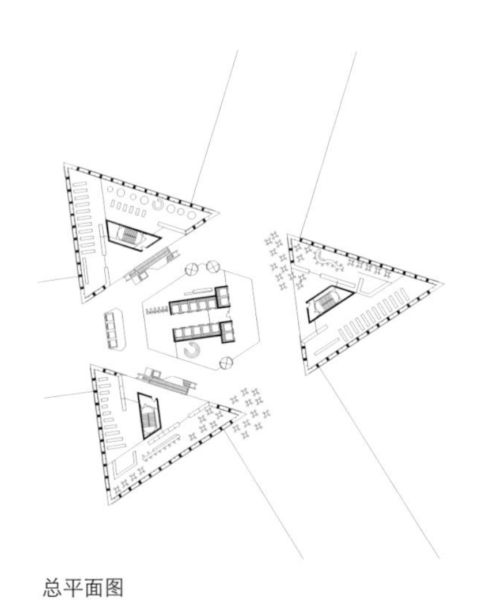
总平面图

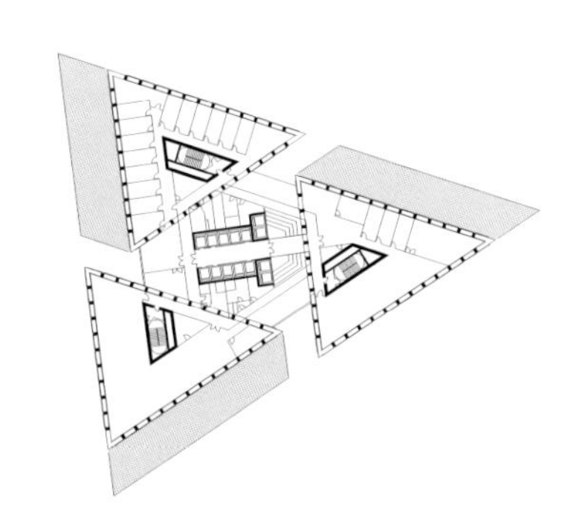
办公室平面图（12层）

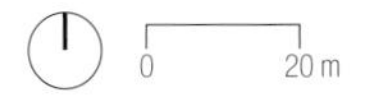

EDF

巴黎脑部与脊椎研究所

巴黎第六大学沙普提厄附属医院脑部和脊椎研究所是法国研究和治疗脑部与脊髓疾病的中心。它是一座建筑面积为22 900平方米的，带有灰色玻璃幕墙的钢筋混凝土梁柱式建筑物。研究所设有实验室、病房、医学影像区，其中还有一半的空间专门作为用于研究实验工作的办公室。教学培训区还包括一个有180个座位的阶梯教室。

该研究所的一个主要目标是以最短的时间将相关研究结果转化到患者的临床应用上面。3座相邻的建筑用于治疗神经系统疾病的会诊与住院设施：分别是神经外科和神经放射科、神经学和神经病理科、精神科。这座8层楼的建筑有两个地下层可以为3座建筑楼层之间提供更为便利的人员流动通路，灵活的基础设施系统也可以满足各个医学研究小组的各种需求。主入口位于H座建筑的两翼之间的凹进处。

威尔莫特有意地规避了很多传统设计，建造了宽阔的透明走廊作为讨论与放松的地方。过去，不同类型的研究在不同区域内进行（包括分子与细胞生物学、神经生理学、认知科学、治疗等），彼此间较为隔绝。而这座大楼的基础部分设计则旨在能够与位于相邻地点的拟建内分泌科和营养学中心实现更好的流通性。该研究所采取的综合学科研究法是医疗研究的一项重要创新。巴黎第六大学沙普提厄附属医院拥有2万多名医疗人员，总计2000张病床，每年可以治疗超过10万名神经系统疾病患者。

项目地点 / 法国巴黎
完工时间 / 2010年
建筑面积 / 22 900平方米

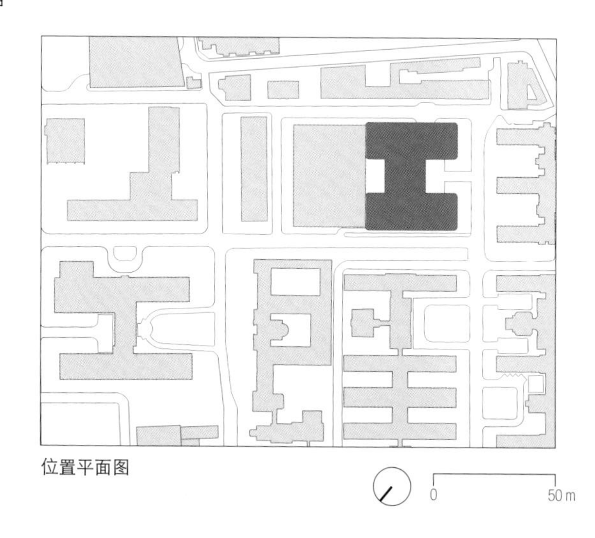

位置平面图

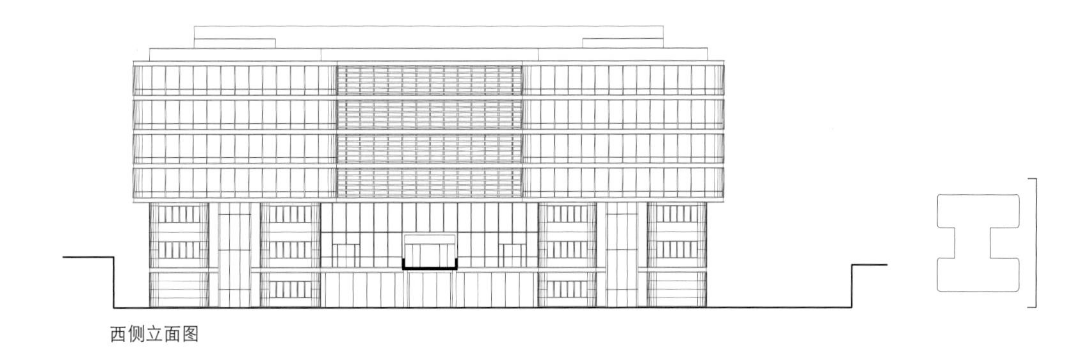
西侧立面图

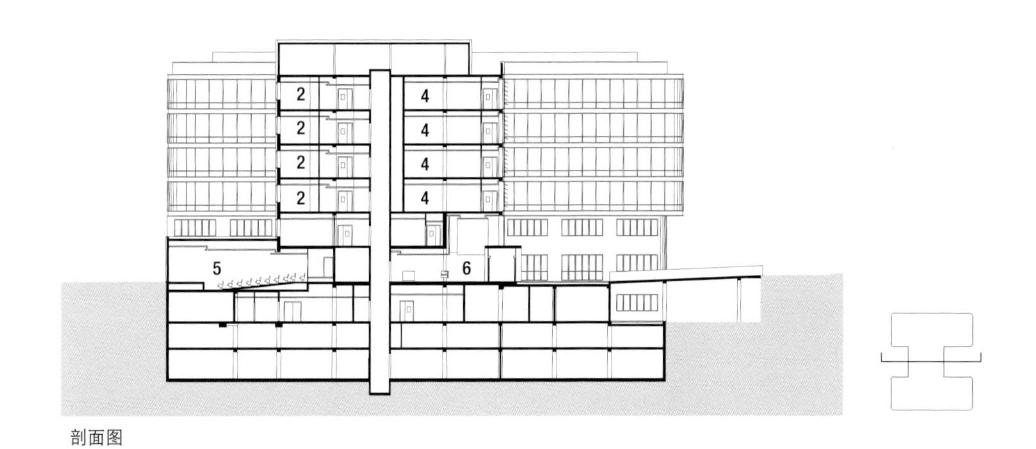

剖面图

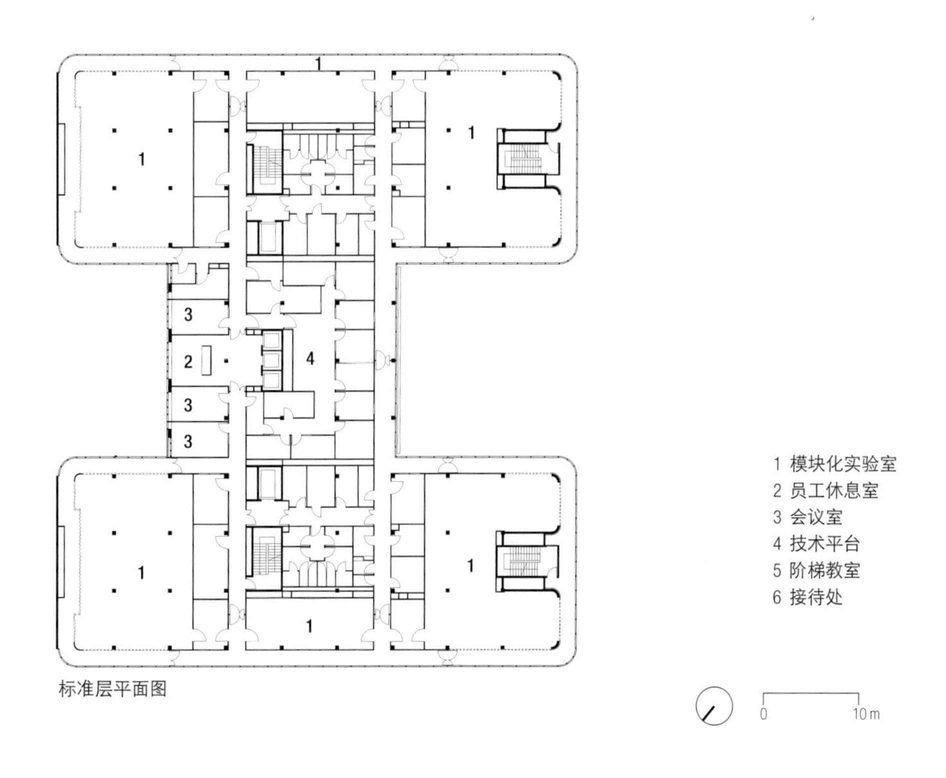

标准层平面图

1 模块化实验室
2 员工休息室
3 会议室
4 技术平台
5 阶梯教室
6 接待处

ICM

ICM

蒙特卡洛
观景塔大厦

这是一座3栋相连的20层公寓建筑，由钢筋混凝土建造而成，外表采用铝蜂窝复合材料幕墙板、不锈钢型材与法帝米黄天然石材等，并且各个公寓住宅设有带玻璃和金属栏杆的凸窗。

大厦的顶层设有两块大型观景窗，突出了建筑的地标性效果，从这里可以看到摩纳哥的东部，同时这儿还给顶层居民提供了一个重要的景观——摩纳哥亲王宫。如此一来，该项目的设计具有了更为贴心的个性表达，从远处也能清晰可见。

大楼下面有一个129个车位的6层停车区域，还有一个零售空间。61米高的蒙特卡洛观景塔大厦位于摩纳哥热带植物园附近，街区周围高楼林立，其建筑面积总计10 000平方米，包括3层楼的办公空间和15层楼的公寓。这座高度现代化的大厦还是尊重了摩纳哥的传统建筑风格（与街道一致）和构成（基础部分与顶楼）。该建筑是为当地著名的地产开发公司米歇尔·帕斯托集团而设计建造的。

项目地点 / 摩纳哥
完工时间 / 2012年
建筑面积 / 10 000平方米

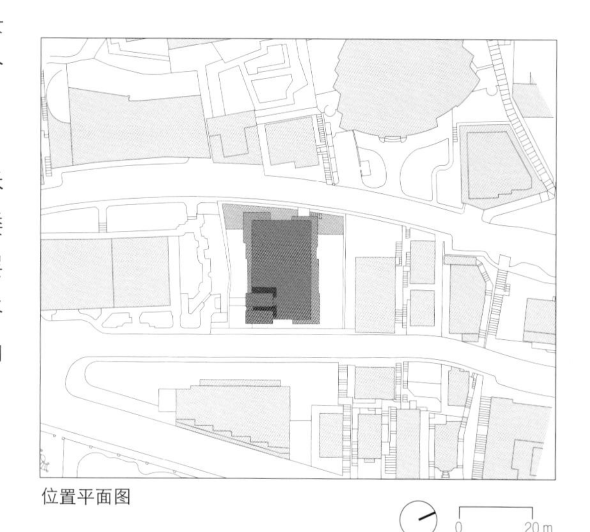

位置平面图

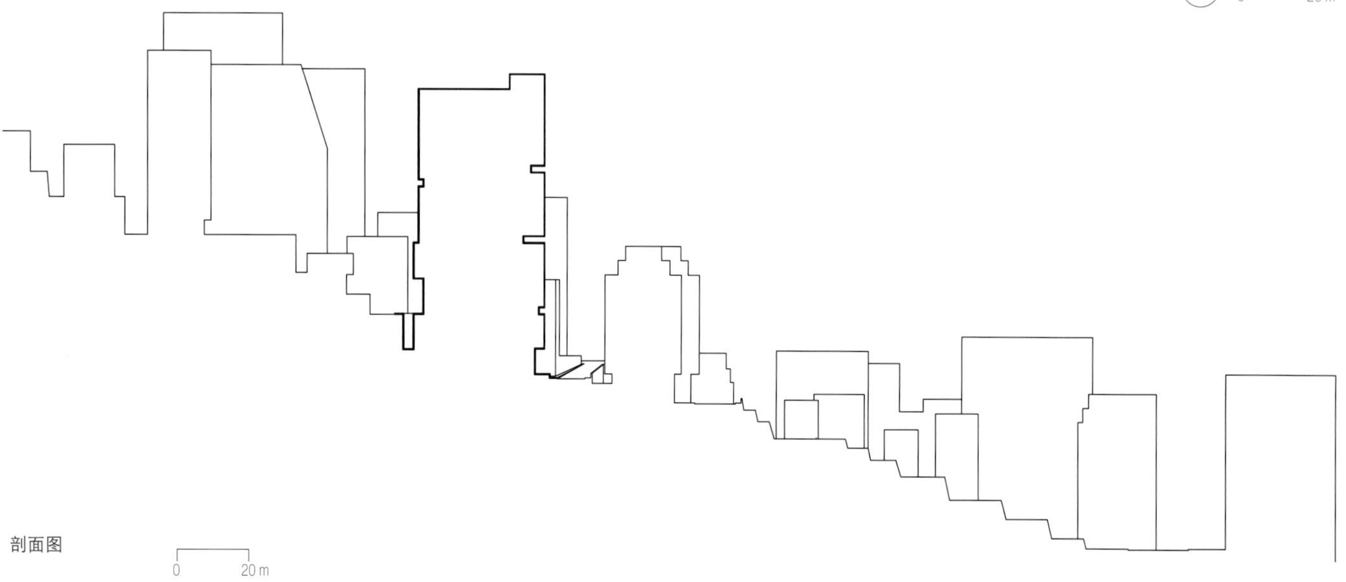

剖面图

1 入口大厅
2 办公室
3 办公室接待处
4 服务设施
5 停车场通道
6 花园
7 卧室
8 客厅
9 饭厅
10 厨房
11 无边泳池
12 日光浴室
13 夏季厨房
14 客厅天花板

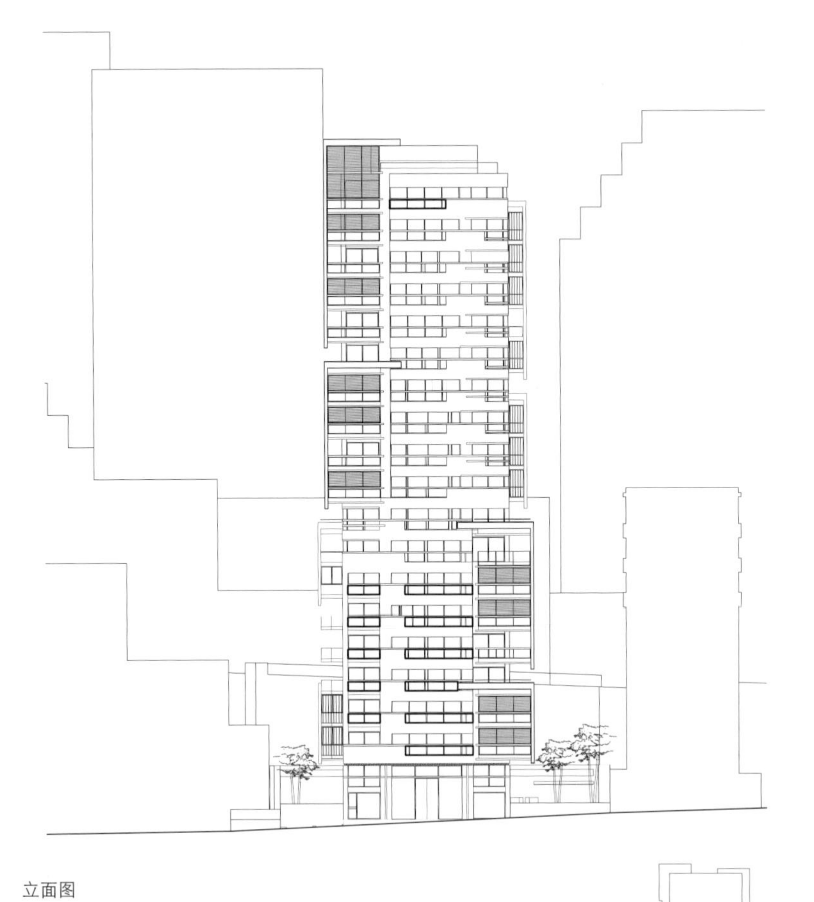

立面图

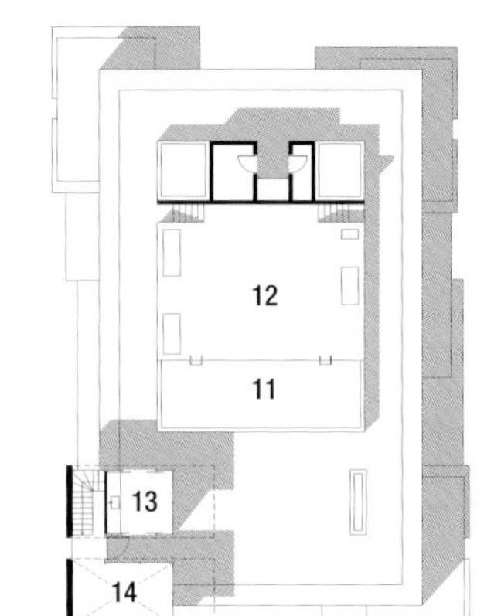

顶楼平台平面图（20层）

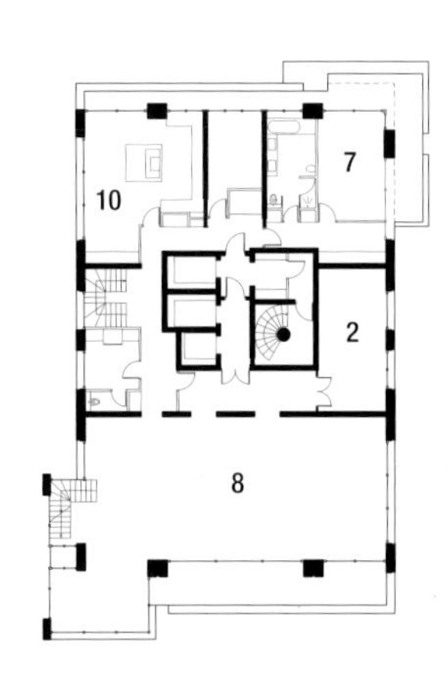

顶层平面图（19层）

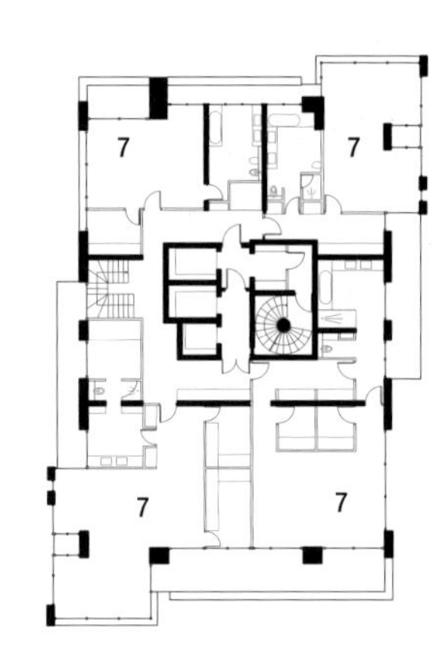

顶层平面图（18层）

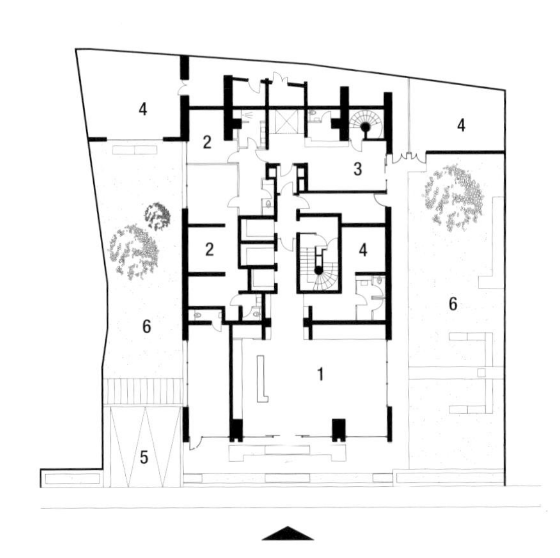

一层平面图

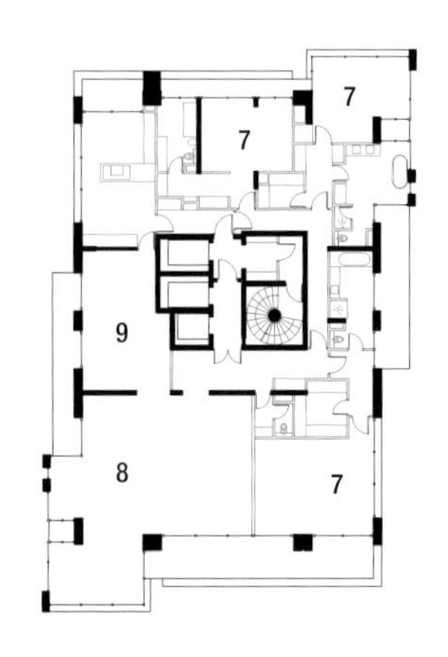

标准层平面图

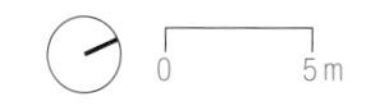

拉德芳斯发展局总部大楼

这个面积达15 000平方米的低能耗办公楼位于楠泰尔十字路口广场，靠近巴黎西部的拉德芳斯商业区。这座钢筋混凝土建造的拉德芳斯发展局总部大楼有着铝制承重板的外立面，可以控制阳光射入的双层玻璃窗和木制地板。

该项目位于楠泰尔特拉斯地区，该地区已依次布局了17个类似的开发项目，使得人们可以从拉德芳斯新凯旋门悠闲漫步到塞纳河。这座建筑的办公空间有八层，一楼有一家对内的餐厅，还有两层地下停车楼层。建设地点的起伏地形使得建筑在各处的高度有所变化，设计师因此采用了一种紧凑的设计，这样既可以充分利用此处的场地，同时又能顾及邻近的项目。

这座建筑看起来像是一块金属巨石，而从它的基座上分离出来一条透明的“玻璃带”在整个建筑体的一楼环绕了近一周。在入口附近，这条“玻璃带”扩大到了三层楼的高度。该建筑的设计严格遵守高质量环境标准，获得了超高能效和低能耗等级评定。设计师在该建筑外立面上所下的功夫在获得这些评级的过程中显得尤为重要。它们在项目的环境影响中发挥了根本性作用，可以将最大量的光线引入内部空间，同时还能将光线反射到相邻的建筑物上。

项目地点 / 法国楠泰尔
完工时间 / 2012年
建筑面积 / 15 000平方米

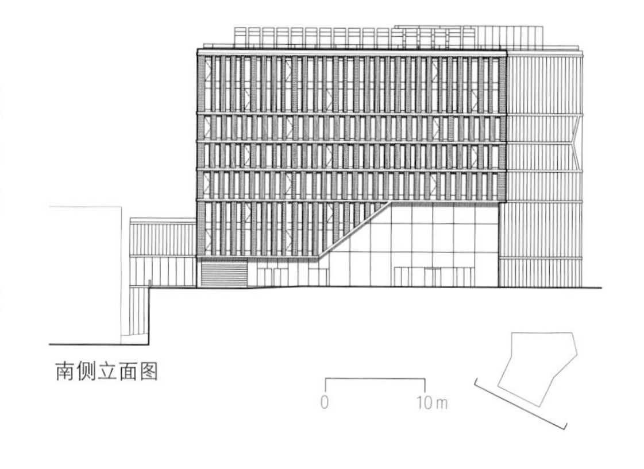

南侧立面图

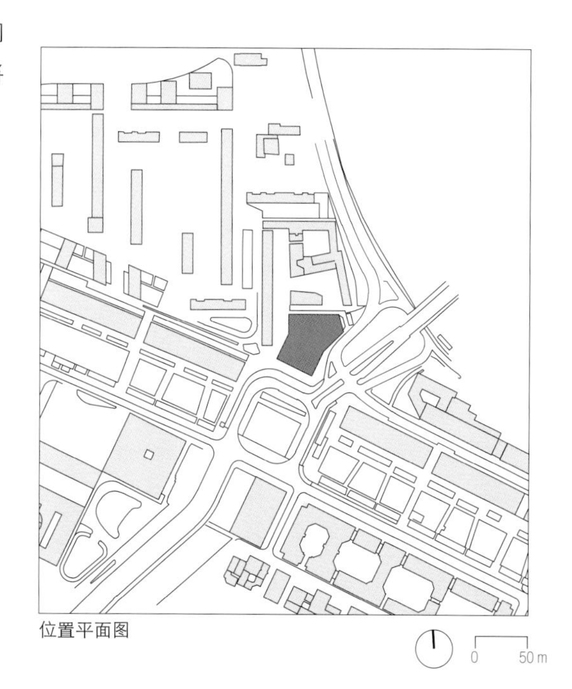

位置平面图

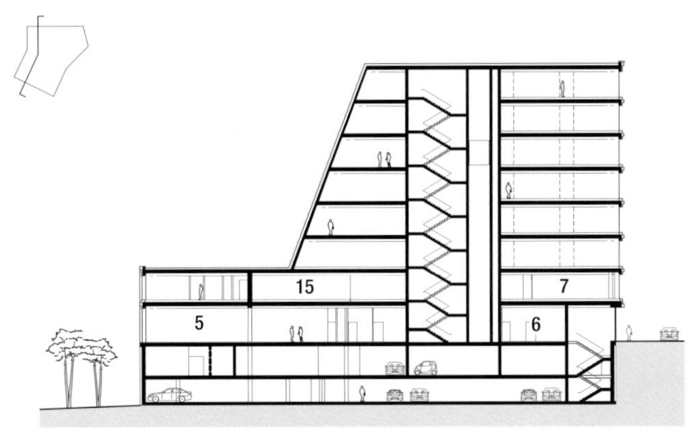

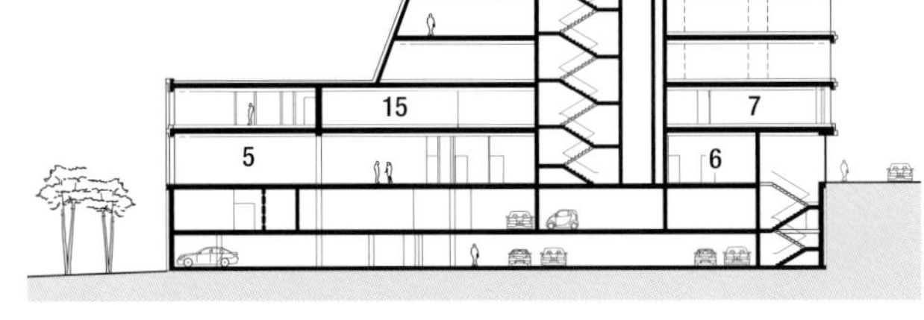

纵向剖面图

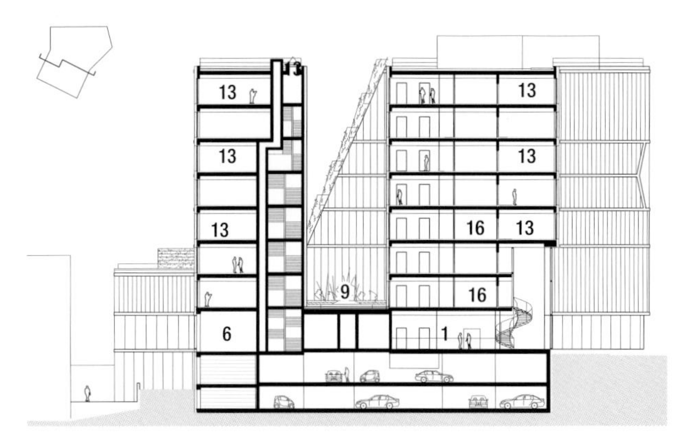

横向剖面图

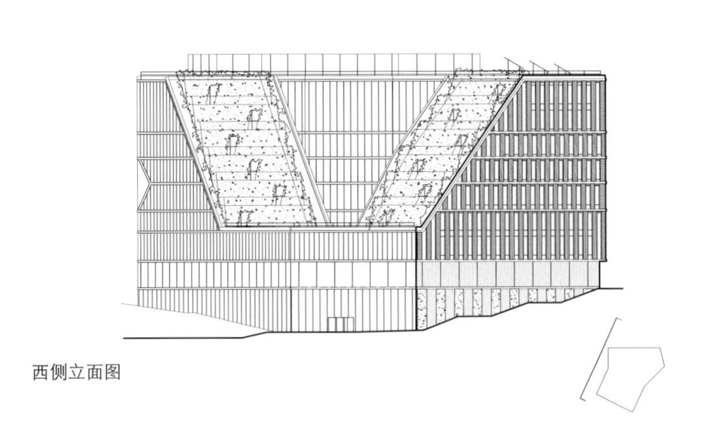

西侧立面图

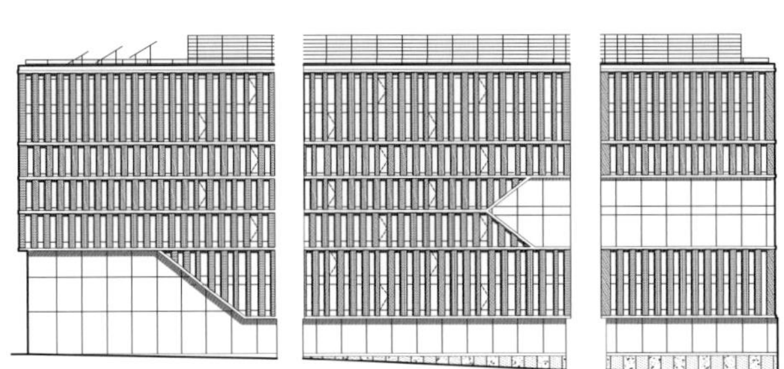
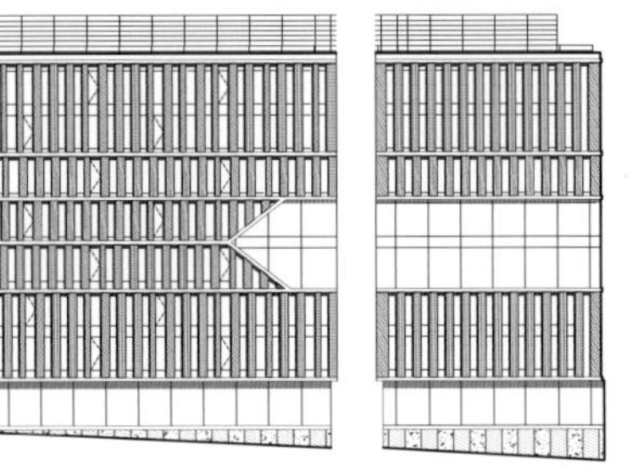
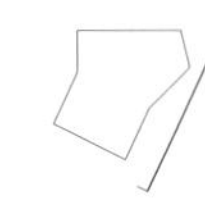

东侧立面图

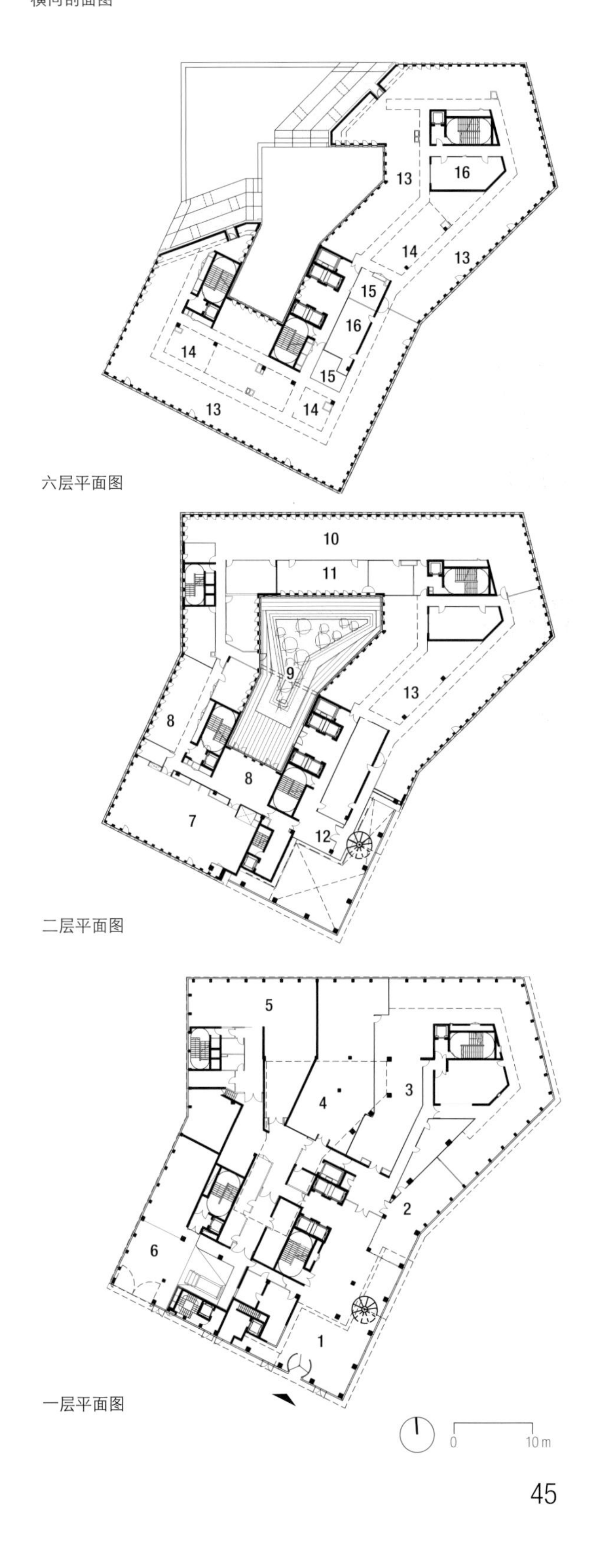

六层平面图

二层平面图

一层平面图

1 大堂
2 咖啡厅
3 餐厅
4 厨房
5 展厅
6 停车场通道/装卸点
7 礼堂
8 大厅
9 平台（花园）
10 商务中心
11 行政人员餐厅
12 商务中心出入口
13 办公室
14 会议室
15 茶室
16 洗手间

NELSON
MANDELA

上诉法院
律师培训学院

该项目是威尔莫特建筑事务所的一个中标项目，巴黎上诉法院律师培训学院是用来为巴黎上诉法院的律师提供专业培训的学校。这座建筑面积为8750平方米的建筑物是由石材、木材、玻璃、钢材、铝材和混凝土构成的，建在一块棕地之上。建筑内还包括一座边长30米的方形中央花园和一个3层楼高的大厅。

这个空间包括一个有360个座位的阶梯教室，并能被夹层楼和阳台俯瞰到。建筑主立面有三面临街，周围有地区快车网C线列车、有轨电车，而且距离河堤路、巴黎环形公路和支路都不太远。该建筑的架构简洁朴素，在某种程度上也是基于当地建筑法规的规定。在建筑上层，外立面采用简单抛光的木制板条，内部则采用双层玻璃窗。

该建筑的可见性因对其反光做出的柔化处理而变得更为突出。这些“双层皮肤（玻璃）”“三重框架（双层玻璃+抛光板）”是由一圈串起来的，2.7米线长的云杉胶合板形成的，它们有一层楼的高度，间距27厘米。没有涂漆的木板可以控制建筑物内的湿度，空气潮湿或干燥时，可以通过每块窗体单元底部的单独开口吸收或释放湿气以调节室内湿度。紫外光灯慢慢转动使得木板变成温暖的蜜色，赋予建筑物令人愉悦的光泽，形成一种动态的光影效果。

项目地点 / 法国伊西莱穆利诺
完工时间 / 2013年
建筑面积 / 8750平方米

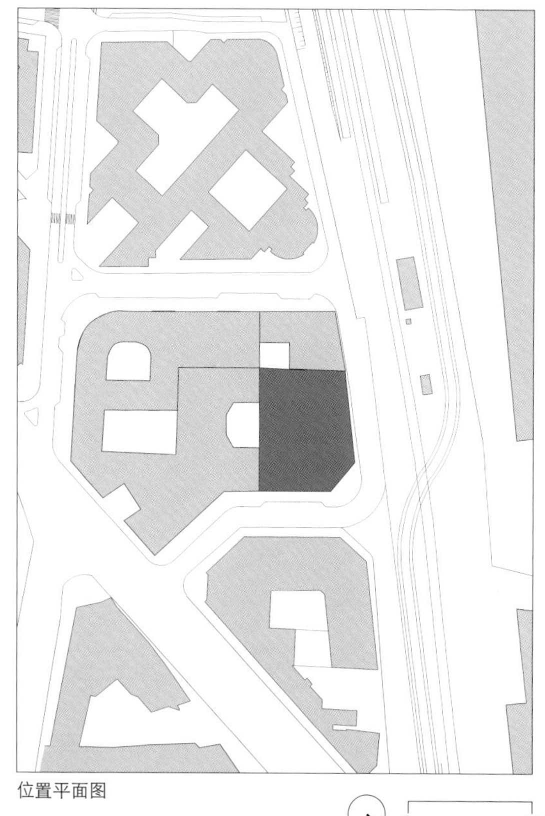

位置平面图

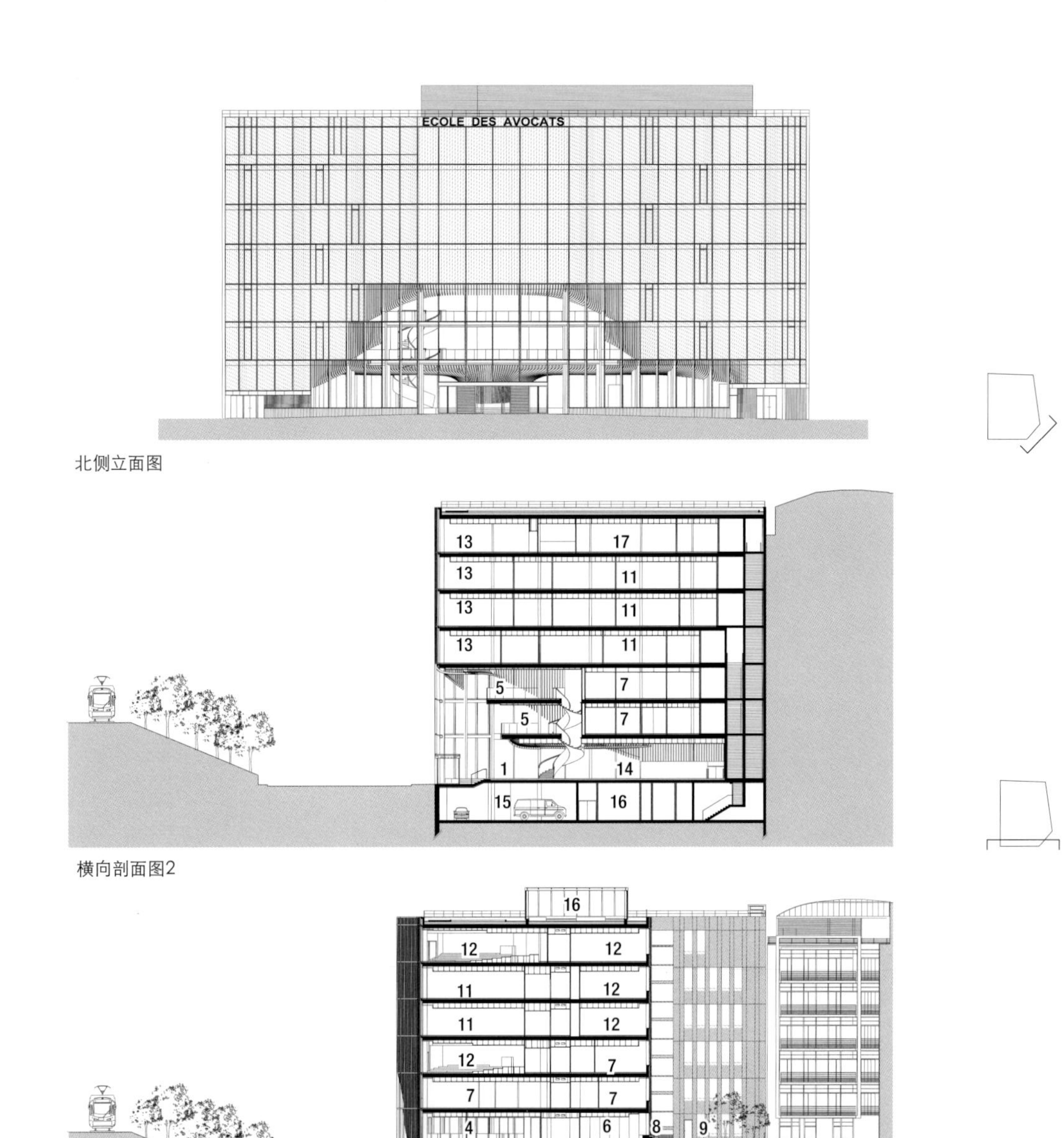

北侧立面图

横向剖面图2

横向剖面图1

1 大厅
2 阶梯教室
3 洗手间
4 图书馆
5 夹层楼
6 咖啡厅
7 办公室
8 阳台
9 花园
10 入口大厅上部空间
11 教室
12 半圆形厅
13 休息室
14 接待处
15 停车场
16 服务设施
17 会议室

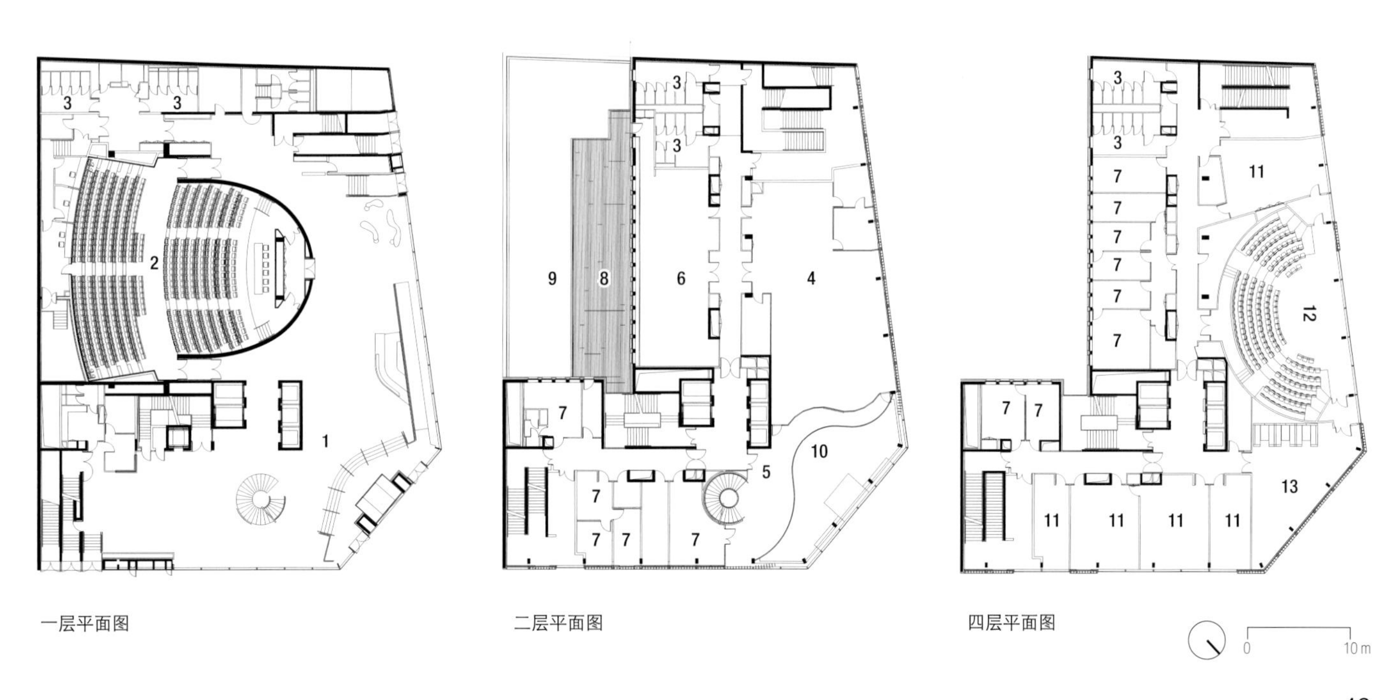

一层平面图

二层平面图

四层平面图

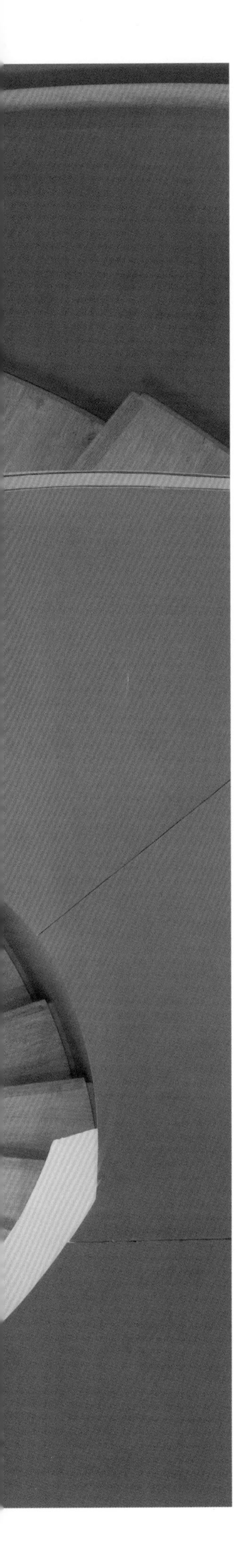

日内瓦国际学校艺术中心

这个艺术中心建有一座有350个座位的多用途礼堂、一座有200个座位的小礼堂、拟向公众开放的展览区、供学生学习与集会的专用空间、模块化教室、排练室、录音与打击乐工作室、多媒体中心和会议室等。该综合体的总建筑面积为4730平方米，主要的建筑材料是混凝土和U形玻璃。

室内设计较为素净，有些材料故意保留了它们的自然状态。为了突出展现学生们的作品，该建筑采用的颜色很少。通道里的中线性照明系统凸显了这些空间的图形化、结构化特征。大厅中使用了不少木质材料，其中橡木质地双层高度墙体标志着礼堂的存在。

该项目旨在将日内瓦的文化艺术活动整合到这所国际学校中来。该建筑是在日内瓦市中心附近的大拉布瓦西埃建造的，并且靠近旧城区。艺术中心项目是日内瓦国际学校大拉布瓦西埃校区改造项目的最后阶段。它将这所综合性学校的所有要素联系了起来，建造出一个统一的活动场所。这座艺术中心还与学校的自助餐厅、体育场及希腊剧院相连。

这座建筑建造出了必要的空间与基础设施，汇集了现有的艺术、戏剧和音乐等教学部门。正如让-米歇尔·威尔莫特作品中最常见的情况一样，他提出了一个强调透明度、强调轻松愉快、强调自然的方案，这些元素在这里被整合在了一起，预示着学生们的创造力与创新精神。

项目地点 / 瑞士日内瓦
完工时间 / 2014年
建筑面积 / 4730平方米

位置平面图

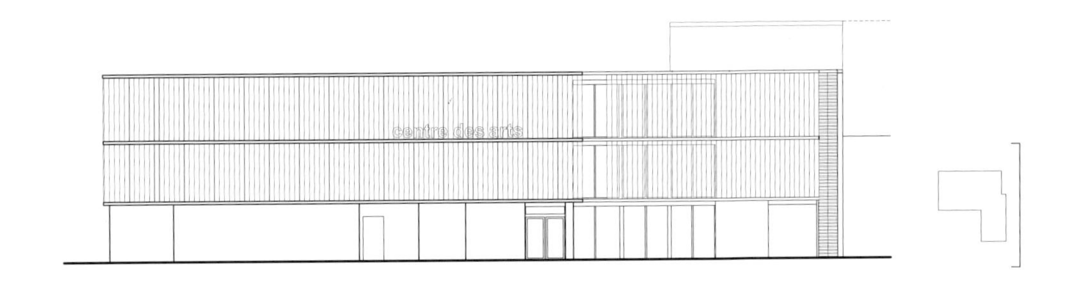

东侧立面图

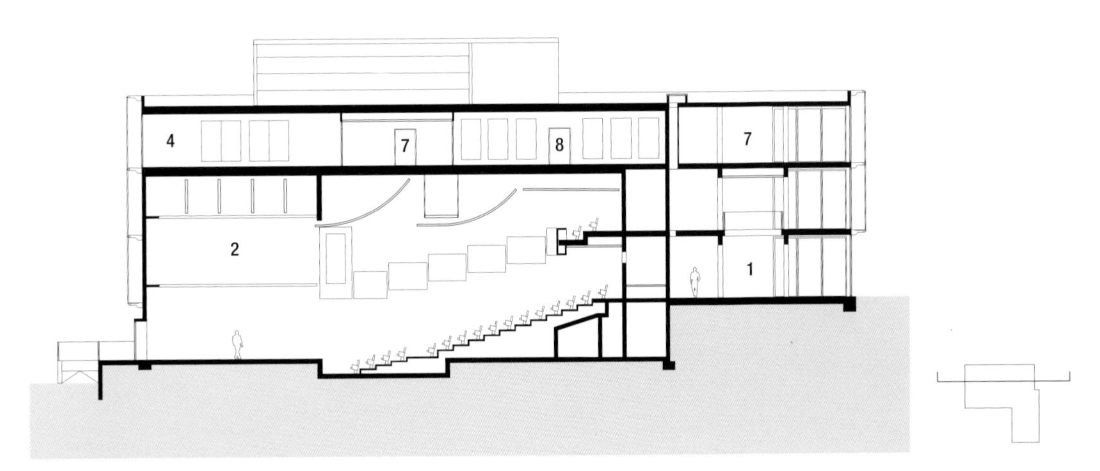

纵向剖面图

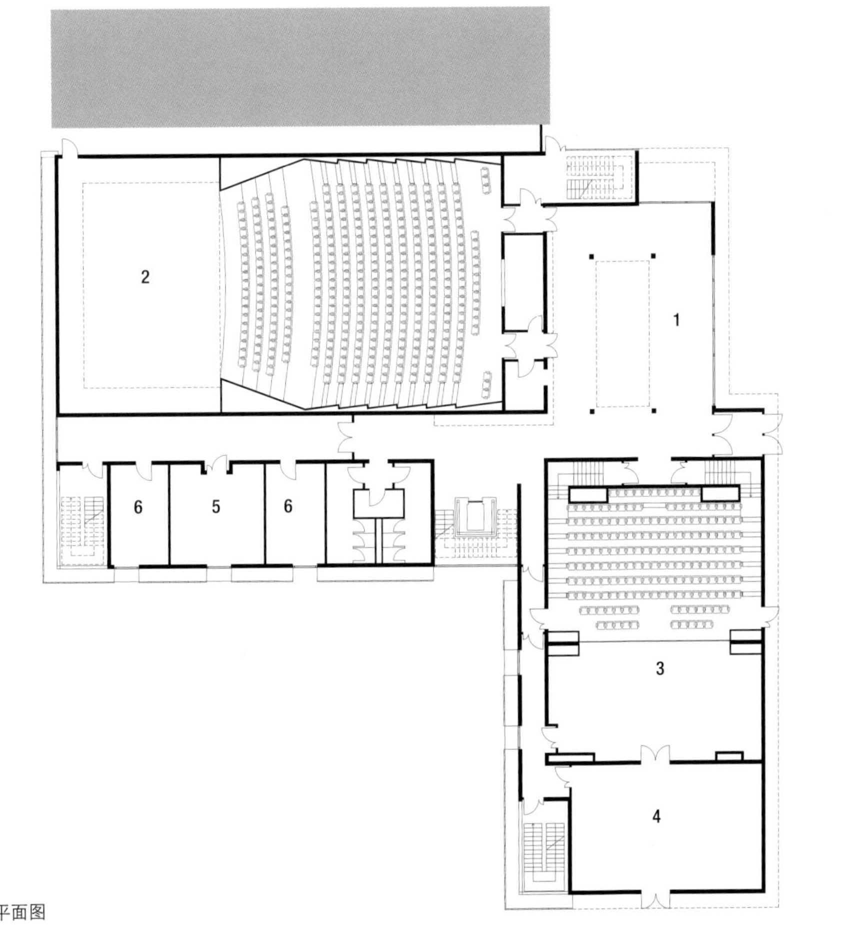

一层平面图

1 大厅
2 礼堂
3 剧院
4 教室
5 会议室
6 办公室
7 学生交流区
8 舞蹈室

0 5m

centre des

arts

安联里维耶拉球场

这座球场位于尼斯的瓦尔平原生态区。该项目包含了一个基于零售、休闲娱乐及餐饮服务的发展计划。该计划旨在打造一个与球场共生共享的、极有凝聚力的综合性设施，使之最大限度地发挥出城市设施的协同作用、用途以及巨大能量。球场项目在一个占地面积14公顷的场地上，有超过其中五分之二的土地被用来进行住宅建设，这座拥有36 180个座位的球场建筑成本为2.17亿欧元。让–米歇尔·威尔莫特和他的团队于2010年中标，并于2011年8月开始施工建造。

这座综合体使用了超过4800个层压板型材和6000个层压板支架，总计消耗了4000立方米的云杉木。亮度与透明度是指导其设计的关键理念。这里没有大的结构性元素干扰球场的中央天孔，使得它可以敞开正对上方的天空。包裹在外立面上的半透明建筑表皮显露出下面飘逸空灵的球场骨架结构，沿着球场表面营造出了一种如梦如幻的光影效果。尽管球场位于地震活动区，但它还是被设计了一个46米高的顶部悬臂结构。球场顶部EFTE（乙烯–四氟乙烯共聚物）膜构建出的起伏形状使建筑给人一种流动之感，看起来就像一大块保护性遮蔽物上扬起来在迎接着观众。

球场内部舒适宜人、采光良好、视野开阔、通道很多。这座体育场旨在为各种活动提供良好的配置，包括体育赛事、音乐会和其他类型的大型公共活动等。

131米×73米的球场可容纳足球、橄榄球、网球、拳击和赛车等体育运动。尽管规模很大，这座球场7500平方米的大锅形状还是在大量坐席和保留给私人使用的接待区中营造出了一种令人惊讶的私密感。安联里维耶拉球场是首先以积极能源策略进行建设的体育场馆之一。屋顶安装的光伏模块覆盖了7500平方米的面积，提供了球场所需的相当部分的电能。球场顶部收集的雨水可以用于冲洗球场和洗手间。通风系统则利用最多风向（盛行风）为内部空间提供自然通风。

项目地点 / 法国尼斯
完工时间 / 2013年
建筑面积 / 14公顷
体育馆 / 面积54 000平方米，座位36 180个

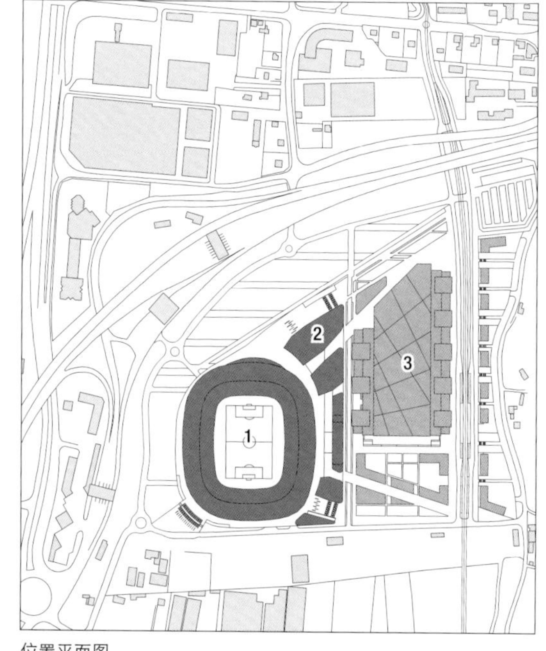

位置平面图

1 球场
2 地产开发办公室
3 与球场相连的生态友好型区域

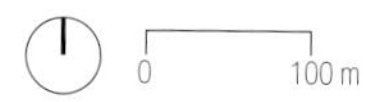

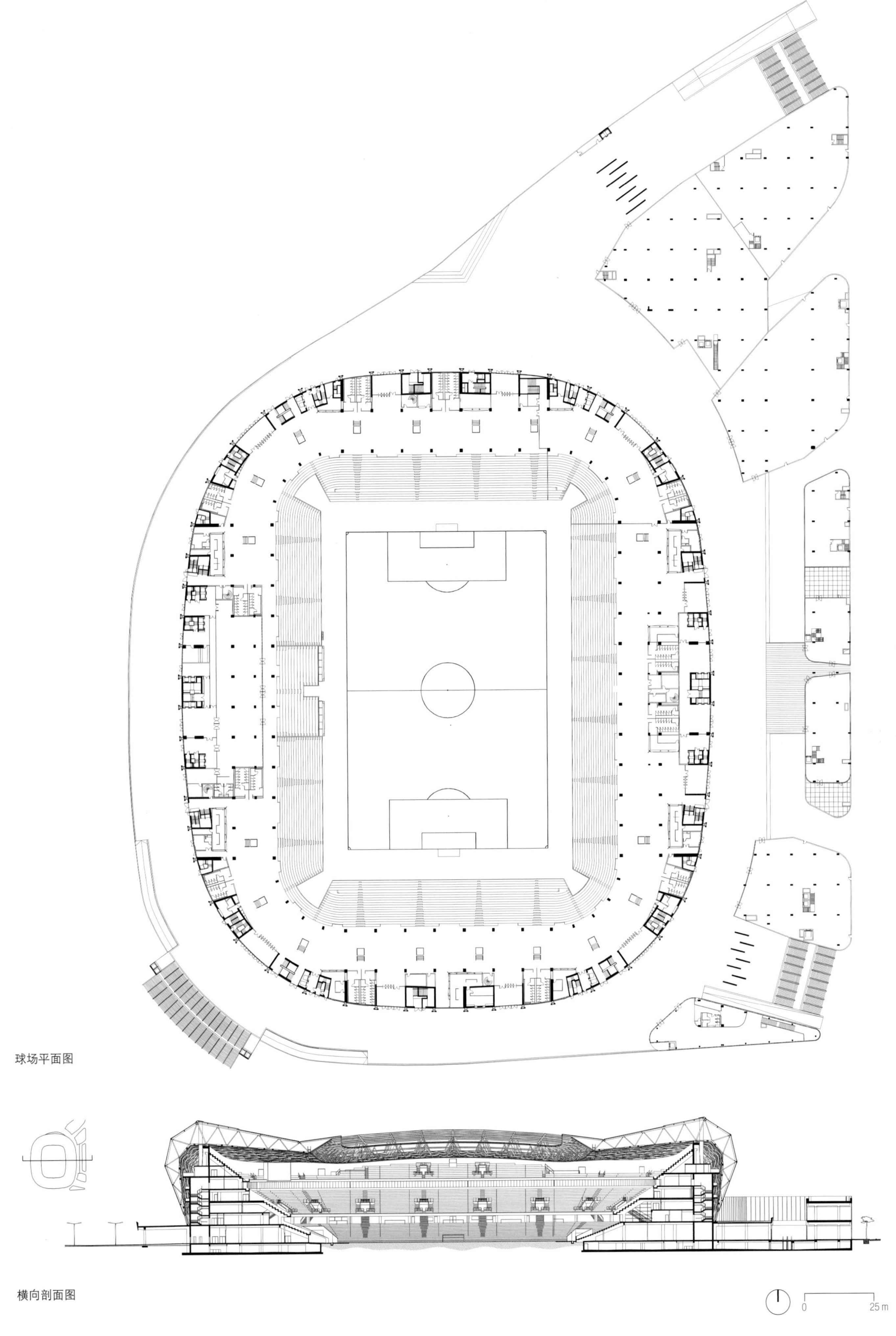

球场平面图

横向剖面图

H1
places
039-078

新国防部大楼
象限角园区建筑

项目地点 / 法国巴黎
完工时间 / 2018年
建筑面积 / 92 200平方米

2015年夏天，让–米歇尔·威尔莫特交付了该建设地点上新法国国防部大楼象限角服务性园区中四座建筑物中的两座。建筑的建设地点毗邻塞纳河，长期被法国军队征用，而且法国空军与海军以前在这里也有相应的建筑。与军方经常采用的方式相反，这个综合体对于城区来说是开放的，而不是像堡垒那样封闭。该项目采用的是巴黎传统的建筑风格，如同大型的、带内部庭院的长方形联排别墅。

这些办公楼占地92 200平方米，占整个象限角园区建筑面积（2.8公顷）的一半，另一半则分配给一个景观花园。“扁平而角度各异的间柱精心安排在外立面上与办公空间中，”威尔莫特解释道，“我们选择的外立面铝材有3种颜色——银色、古铜色和锰黑色，这样当日影移动时，建筑的外墙在日光下会变得富有生气。”

整个项目包括4座7层楼的建筑和景观公园。它们都是钢筋混凝土柱梁结构。所用的主要材料是混凝土、石材、不锈钢、考顿钢、复合板、电镀铝框架玻璃幕墙等，还采用南美蚁木建造了阳台。双层高度的入口大厅和全玻璃的上层楼面共同构成了一种良好利用自然光的整体策略。

阳台和屋顶种下的大量绿色植物与440平方米的太阳能电池板为这些建筑提供了能源需求中很大的一部分。而各个办公室都可以调整大小，可以提供最大程度的内部适应性。

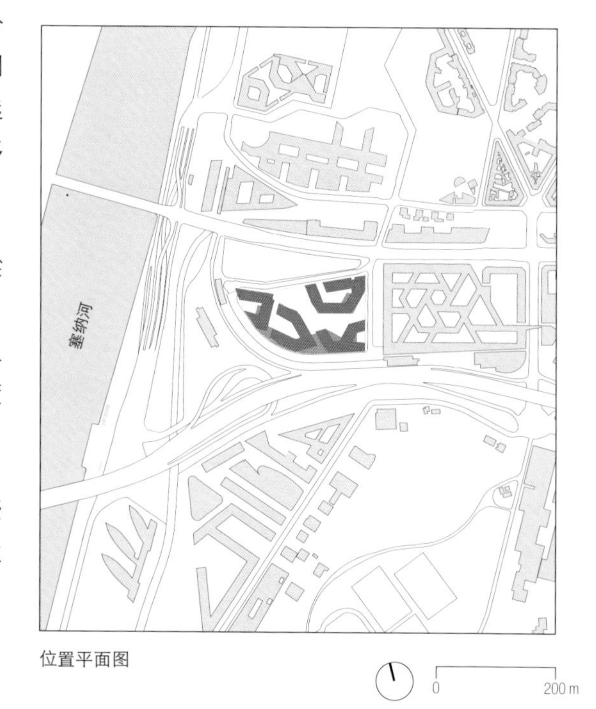

位置平面图

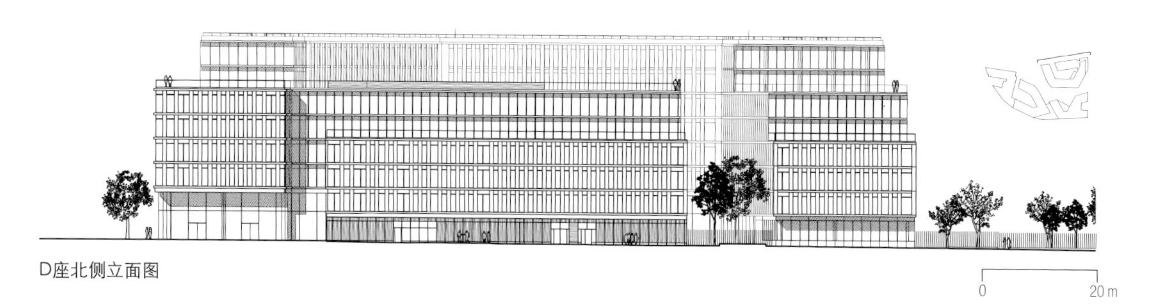

D座北侧立面图

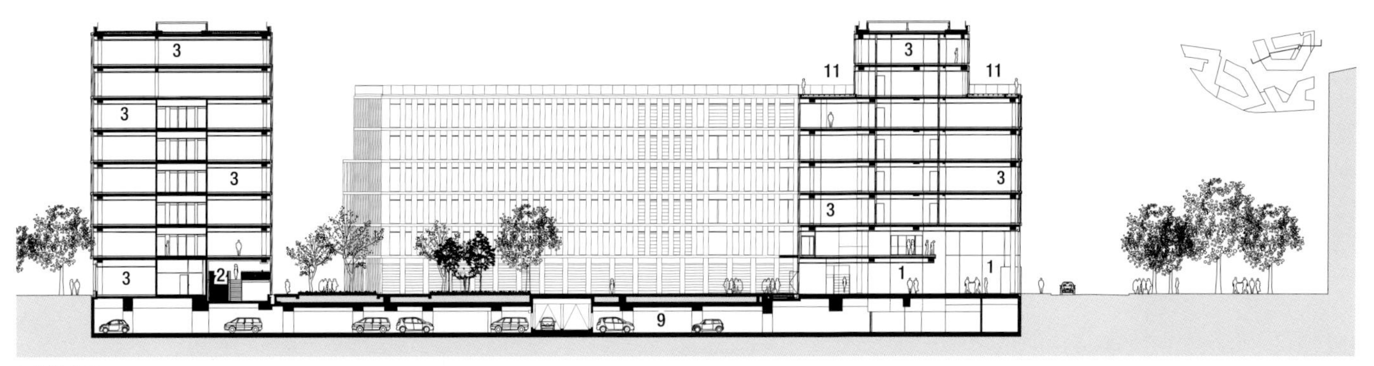

D座横向剖面图

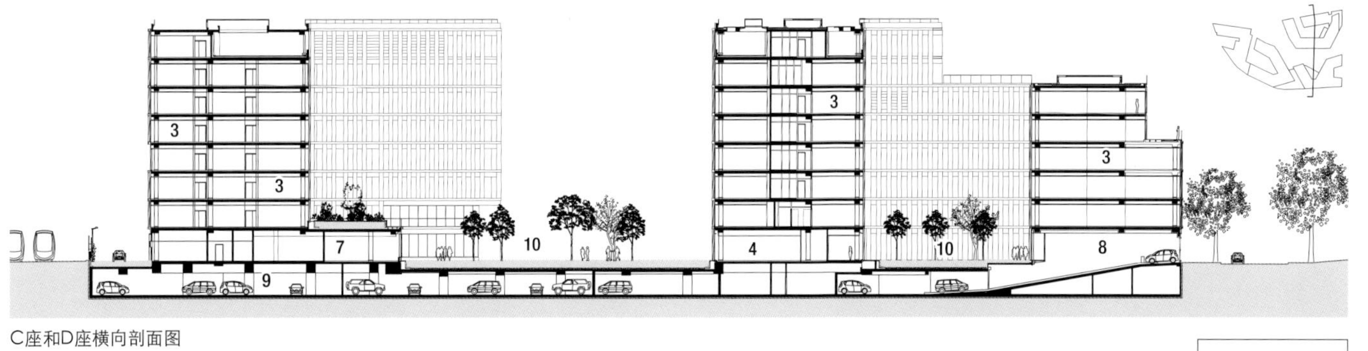

C座和D座横向剖面图

0　　20 m

1 入口大厅
2 单车室
3 办公室
4 会议室
5 咖啡厅
6 商店
7 餐厅
8 停车场通道
9 停车场
10 花园
11 阳台

总平面图

0　　20 m

潘克拉斯6号方形大楼

这个项目位于圣潘克拉斯国际车站（欧洲之星列车终点站）对面，为法国巴黎银行房地产公司所建造。这座混合用途节能型办公大楼是在国王十字车站中心区南部打造的第五开发区里面的六个建筑之一，因此，符合该区域的总体规划。该建筑正在按照英国建筑研究院绿色建筑评估体系的BREEAM2008新标准进行评估，并有望被评为“优秀”级。这是一座建筑面积为49 400平方米的钢筋混凝土（地下室和地面部分）与钢结构（上层采用金属楼板）建筑。建筑使用的主要材料是陶瓦、电镀铝材、混凝土和玻璃等。让-米歇尔·威尔莫特还同时负责了建筑物内部公共区域的设计。

该项目的旁边是国王十字车站和圣潘克拉斯车站这样的一级历史建筑，南边有二级历史建筑“德国体育馆”餐厅和斯坦利大厦等。由于周围的建筑结实耐用、结构简单，所以本项目材料的选择与外墙的细节也取决于周围这些历史建筑的传统风格与特征。项目的外墙是以带陶瓦间柱的钢结构建造的，营造出一种华贵之感，并与玻璃部分形成了鲜明的对比。这个幕墙系统以其相互交织的金属框架和陶瓦间柱对附近的砖石建筑进行了重新诠释。

这个方案的构思是基于要创建一个带有中庭而两侧有两个核心区的设想，这样可以从结构上释放建筑结构与立面的内部楼面空间，打破建筑内部的限制。这种结构基于一种水平与垂直的层次。较低的两个楼层做了调整，为建筑物提供坚实的基础，使之对于公共区域来说更为高效、更为直接。在此之上，中间的楼层和顶楼由一种3米宽的网格系统构成，以实现一些该建筑的功能特性。可持续发展从一开始就是设计构思的一个组成部分。该项目融合了被动设计、高效能建筑服务设施与低碳型能源供应的各种特点。

项目地点 / 英国伦敦
完工时间 / 2015年
建筑面积 / 49 400平方米

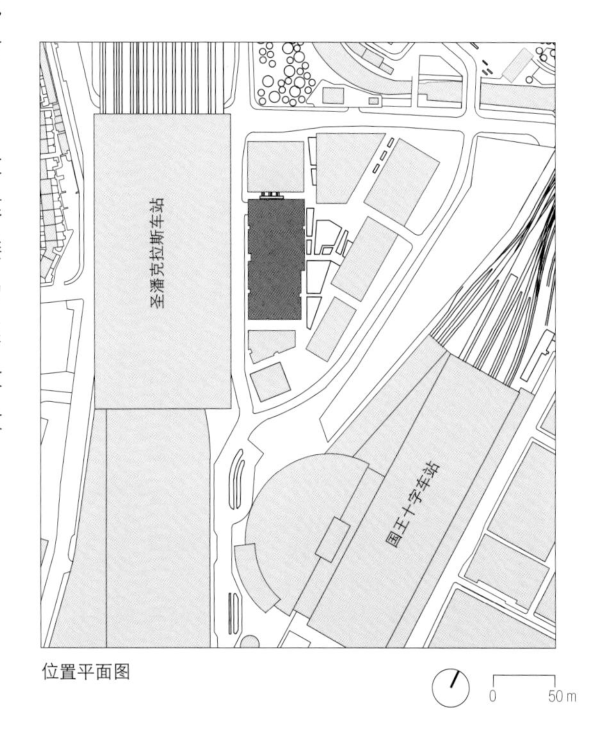

位置平面图

1 主入口
2 中庭
3 办公室
4 零售点
5 咖啡厅
6 洗手间
7 通往地下室的坡道
8 阳台
9 服务设施

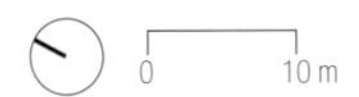

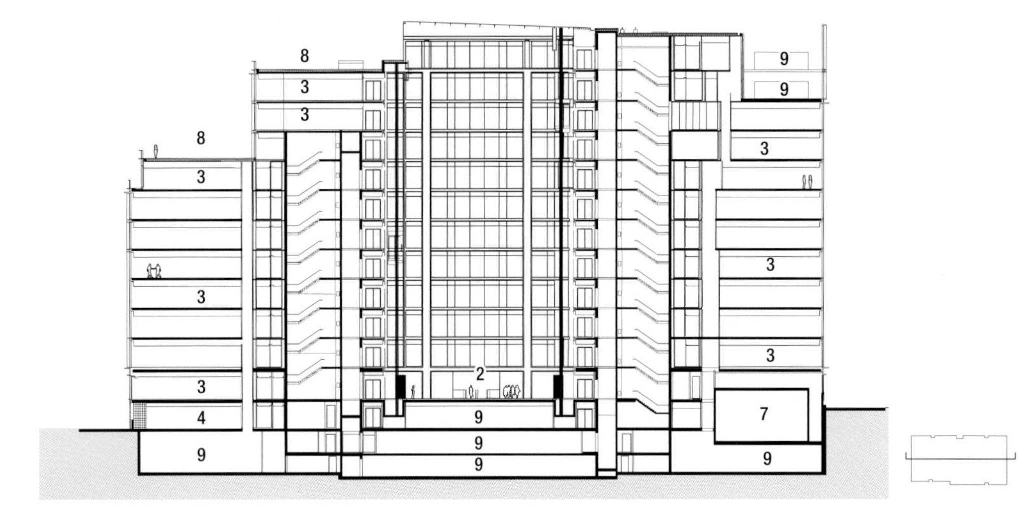

纵向剖面图

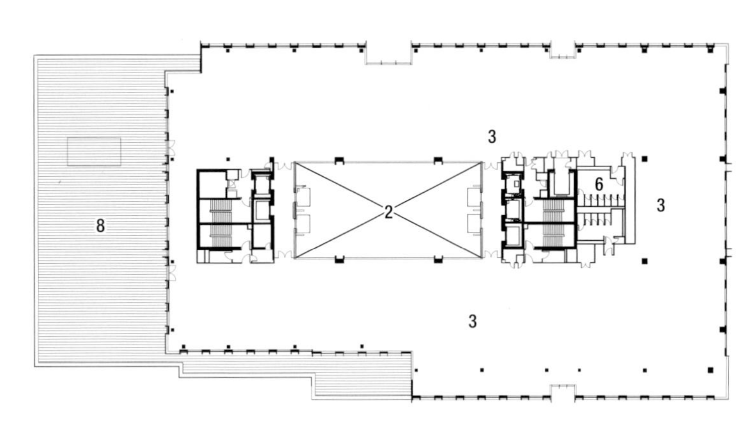

九层平面图

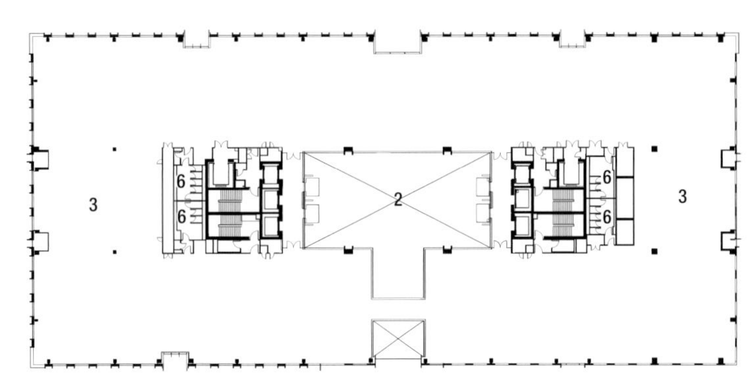

二层平面图

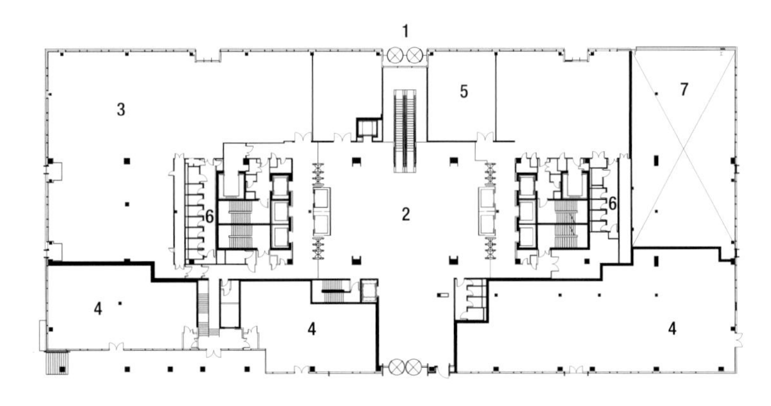

一层平面图

GATE
8

大田文化中心

本项目位于韩国大田市中心的一块三角形地带，大田文化中心是一栋9130平方米，外覆层是铝板和玻璃的圆形钢筋混凝土建筑。建筑内部建有一座350个座位的剧院、一个展厅、一间音乐工作室和一个有79个车位的停车场。这座建筑的每个楼层都包含项目规划中的一种要素，从最具公众性的楼层到最具私密性的楼层，自下而上依次上升。

椭圆形多功能剧院是大田文化中心的核心，位于第三层的展览区与大厅通过两个巨大的楼梯相连，围绕建筑的外部有一个雕塑公园。与这些艺术工作室相关的项目被分散到整个建筑体中，自然光沿着建筑内墙照射进来，人们可以欣赏到雕塑公园的景色。办公室设在第五层（上面有一个屋顶花园）。一个大型的工作室空间位于东立面的“凸点”位置。

尽管文化中心周围有着各种城市开发项目，但它还是能从三个方向被人们看到，这个中心作为一个地标性建筑从周围脱颖而出，成为大田市重要的文化标志性项目，各种文化艺术活动在此上演。在东面，建筑向前延伸，就像一朵向日葵，从而营造出一个从远处就可以看到的巨大的入口立面。建筑物在场地内的排列布置对于公共空间来说产生了两种不同的效果：东面，有一个通过连续的楼梯进出的大型三角形广场，面向城市方向；而西面的空间则更为私密，它位于竹林和中心西侧外立面之间。玻璃幕墙使得展品可以从外部被看到，同时也在白天为建筑带来了自然光线。

项目地点 / 韩国大田
完工日期 / 2015年
建筑面积 / 9130平方米

位置平面图

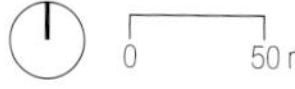

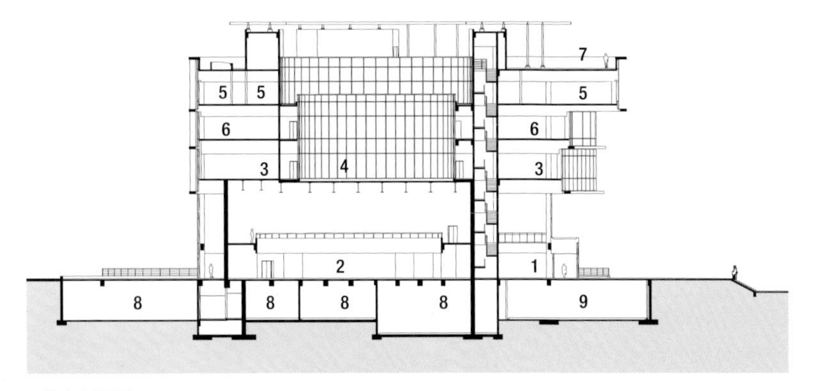

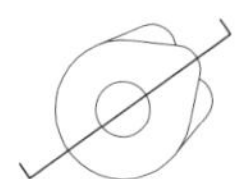

纵向剖面图

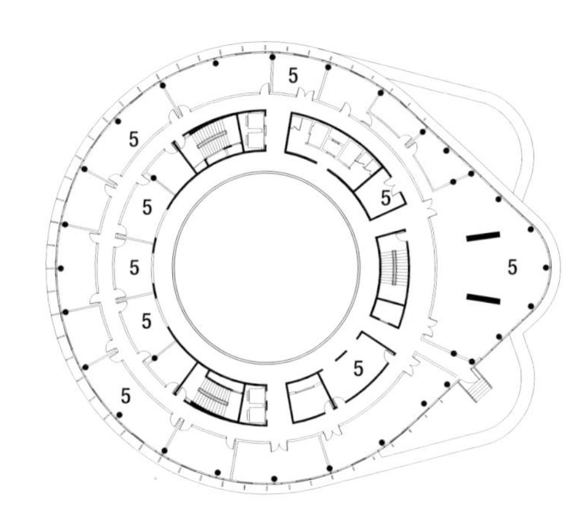

六层平面图

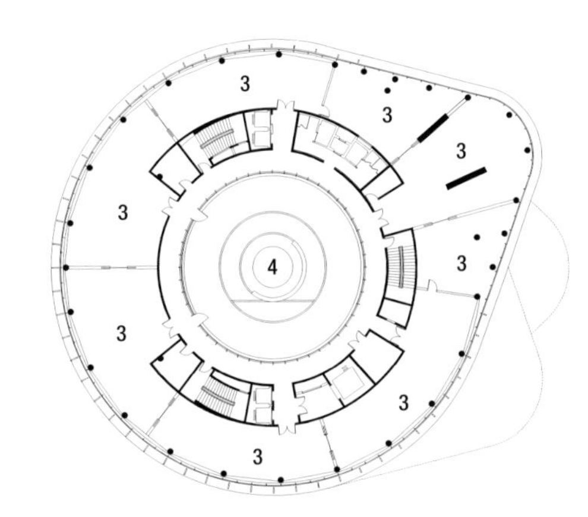

四层平面图

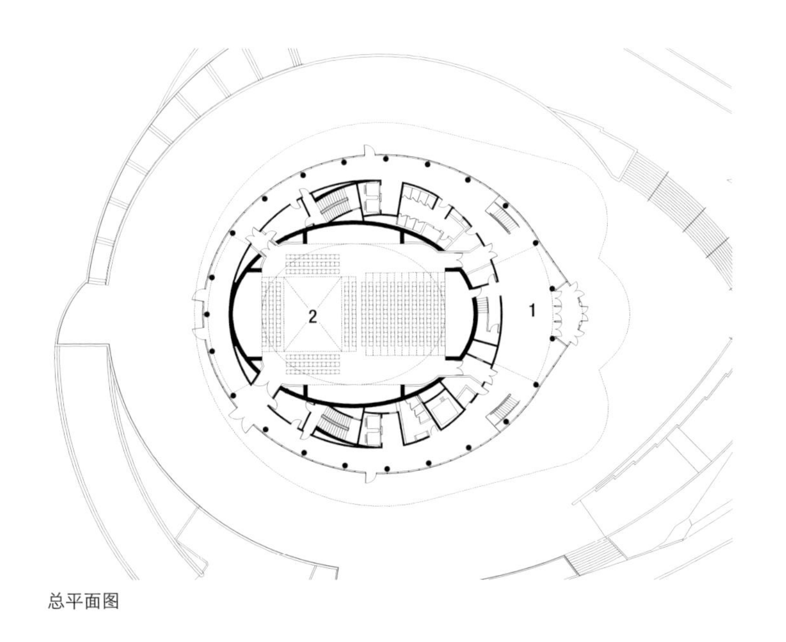

1 大厅
2 多功能室
3 展览区
4 庭院
5 办公室
6 工作室
7 屋顶平台
8 服务设施
9 停车场

总平面图

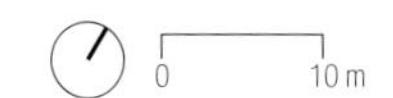

안마
임대
672-
4047
주사랑교회

대전예술가의집

24시 사우나

法拉利运动跑车管理中心

致力于法拉利一级方程式赛车的设计与组装的运动跑车管理中心是一座占地面积24 000平方米的钢筋混凝土建筑。它的外立面是丝印玻璃嵌板（底层）和双层玻璃（上层）。这座建筑的各种元素都是源于法拉利要将之打造成其品牌标志性建筑的愿望，因此具有强烈的辨识性，它将在未来几年与法拉利车队一起发展演变。

室内的赛车设计部门有三层，包括设计师区域、技术人员办公区域、研发区域、交流区域和管理人员区域等。这座建筑中有三个主要的共享空间：运动跑车管理中心员工区、外来访客区和材料与产品区。这个项目面临的关键挑战之一是将这些空间以高效协调的布局组织起来。法拉利标志性的红色、黑色和金色以及建筑表皮的条带环绕形状，都有助于营造出建筑物的力量、特色，给人一种跑车般的运动之感。建筑将法拉利品牌特征中的速度、性能和加速度等这些内在的元素转化为一种形象的空间诠释，使得这座建筑转变成一个被玻璃包裹的“容器”，其建筑外立面不仅可以控制自然光的射入，同时也有效地隔绝了从外部对跑车中心私密区域的窥探，而保密性也正是这种高性能跑车中心的基本条件之一。

屋顶安装了太阳能板，既能提供热能，又能进行光伏发电，同时还安装了雨水回收系统由压缩机对建筑供水，这个设计使得建筑的绿色植物和景观都能很好地融入此地，并达到环保与可持续发展的标准。

项目地点 / 意大利马拉内罗
完工时间 / 2015年
建筑面积 / 24 000平方米

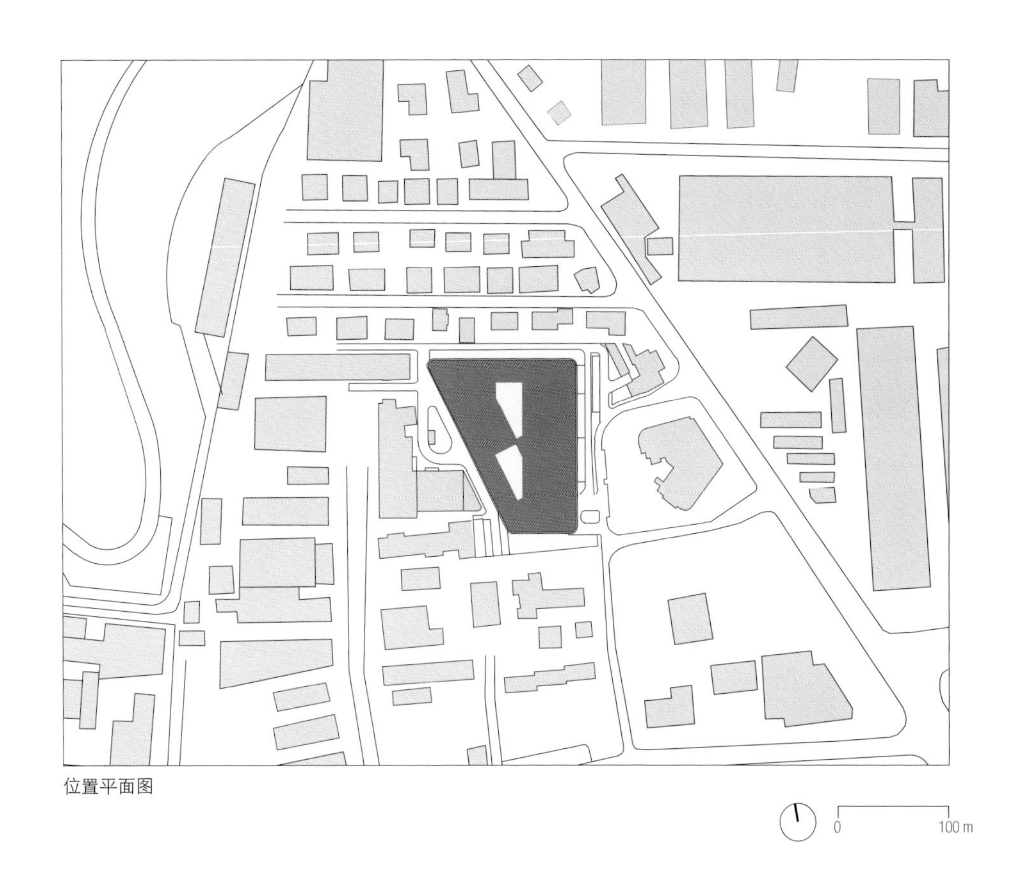

位置平面图

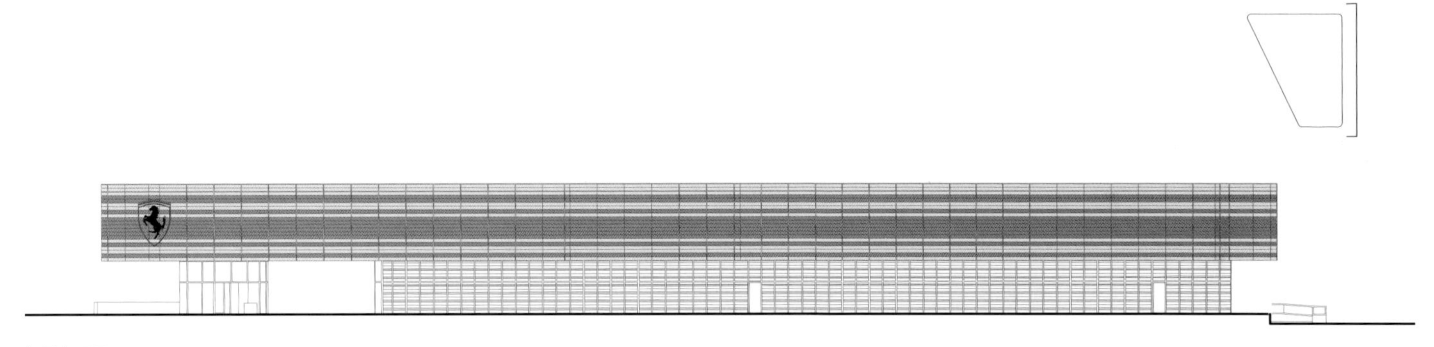

东侧立面图

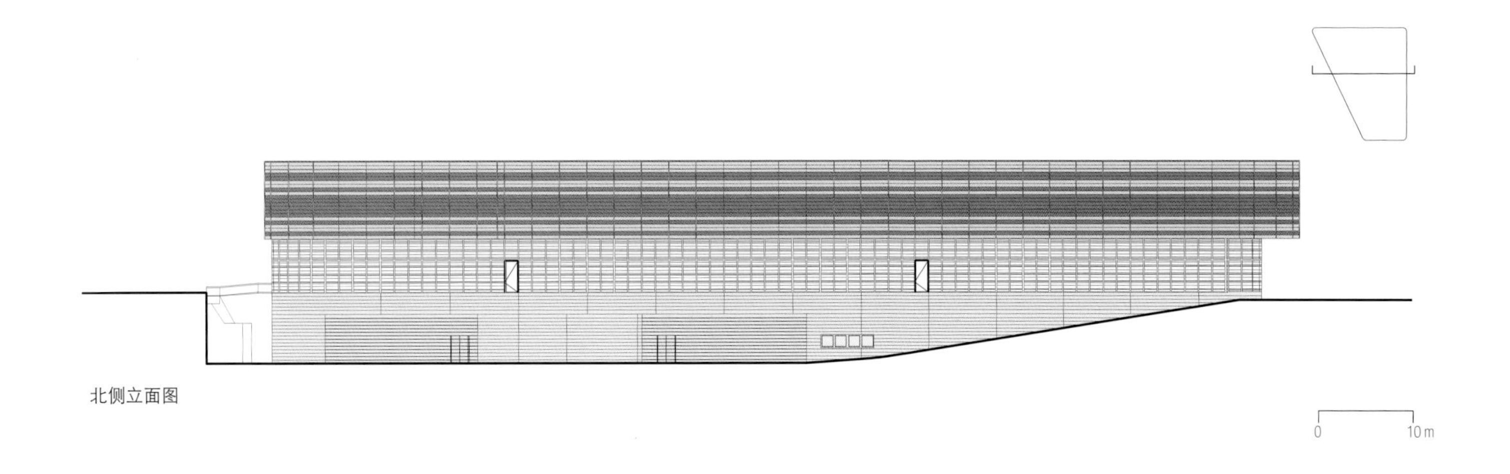

北侧立面图

俄罗斯东正教精神文化中心

巴黎的俄罗斯东正教精神文化中心坐落在塞纳河畔，离阿尔玛桥和布朗利河岸博物馆不远。它建在法国国家气象服务中心大楼的旧址上，靠近塞纳河岸的各大银行建筑群。

实际上，这座建筑物中有四座不同的建筑，分别是一座东正教教堂、一个教区大厅和宗教中心、一所学校和一个文化中心。教堂顶部是五个东正教特有的镀金圆顶，上面竖立着俄罗斯东正教十字架。这些圆顶十分突出，使该中心成为塞纳河左岸天际线的地标性建筑。

在设计这个综合体时，建筑师遵循了三个主要原则：尊重俄罗斯东正教的教规；将之融入巴黎典型的密集建筑云集的街区之中；期望能建造出一座非常现代的建筑。受到那些颇为严肃的中世纪东正教教堂的启发，建筑师采用了一种极简方式，并选用了与附近夏乐宫和埃菲尔铁塔基座同样的石材。在水平方向上，这种石材以断层样式精心覆盖在经过仔细计算的简洁建筑体之上。这座教堂的五个金色穹顶占据了整个建筑综合体的中心，极具辨识度。

这座面积为4655平方米的钢筋混凝土结构建筑是为俄罗斯联邦政府建造的，它使用了马桑吉斯天然石材（法国砂岩）、玻璃和金箔等材料。

项目地点 / 法国巴黎
完工时间 / 2016 年
建筑面积 / 4655平方米

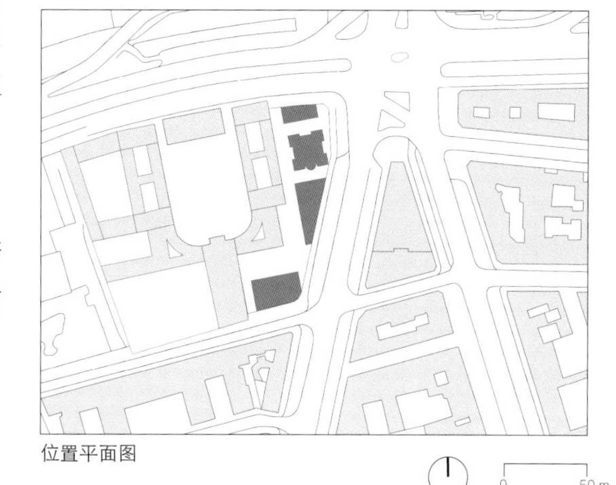

位置平面图

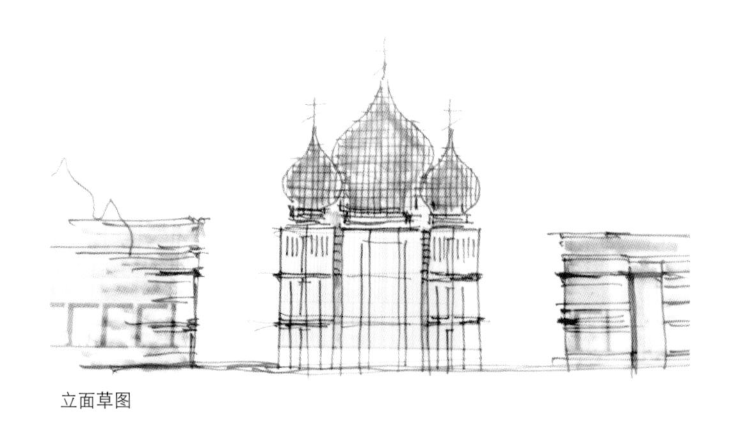
立面草图

北侧立面手绘图

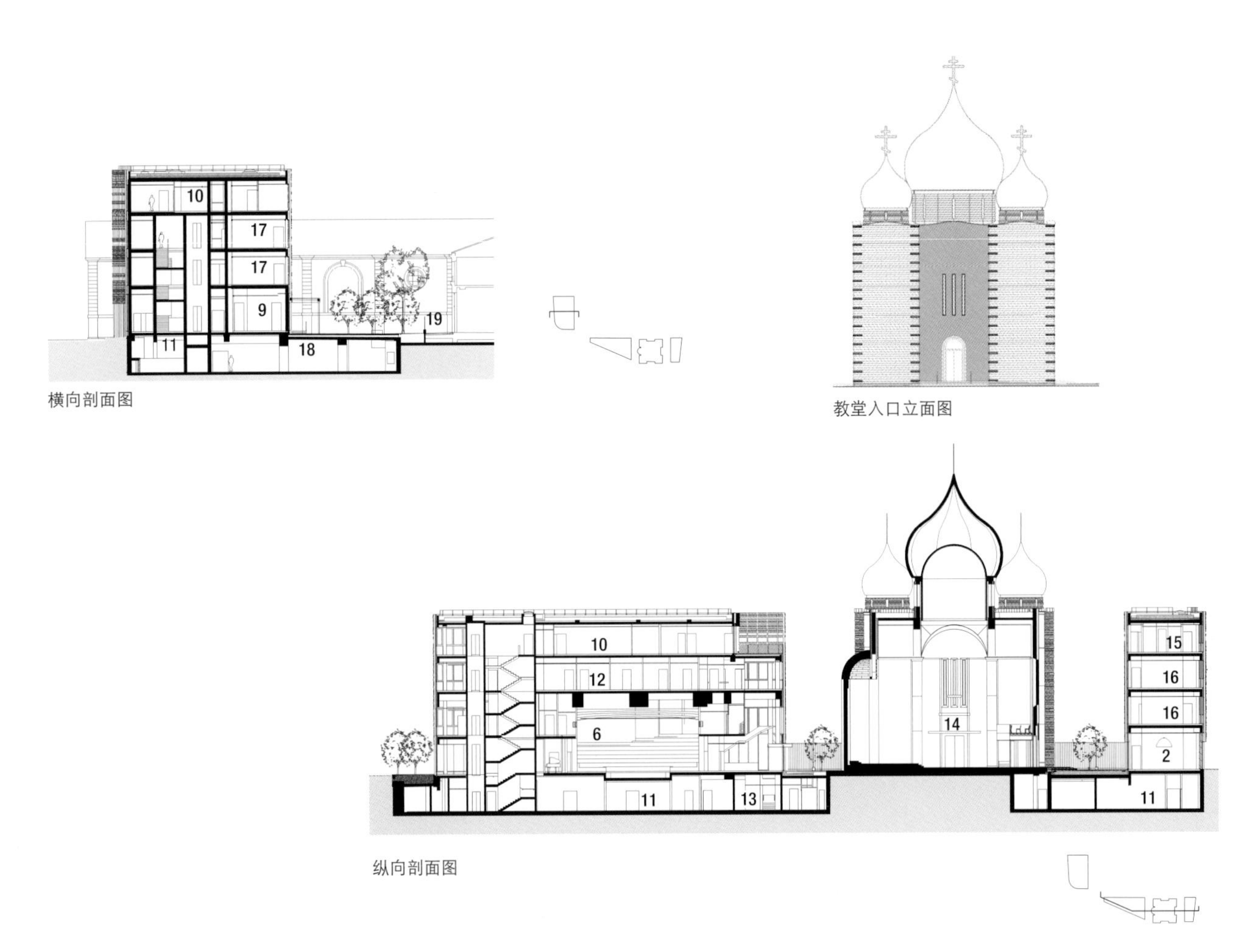

横向剖面图

教堂入口立面图

纵向剖面图

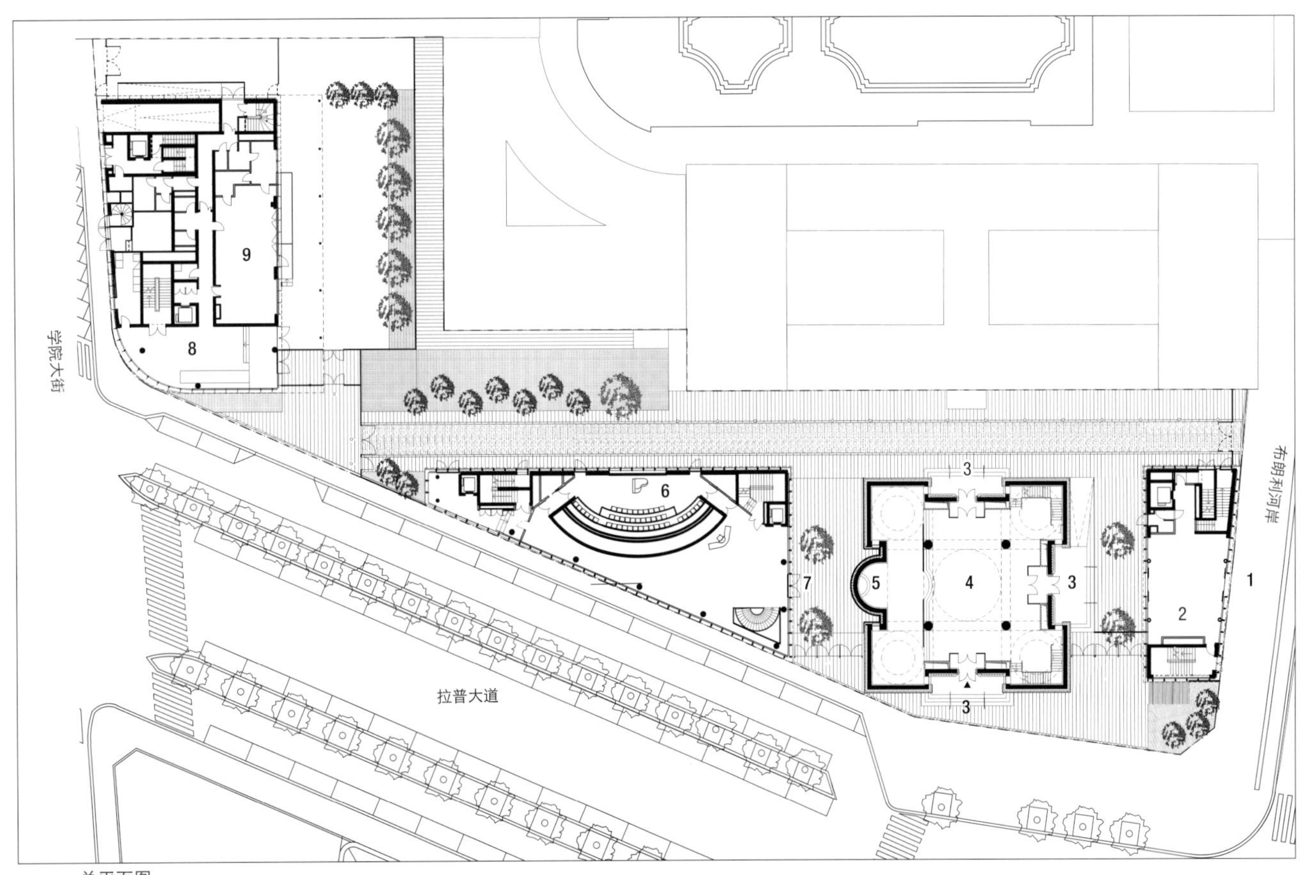

总平面图

1 布朗利河岸的东正教中心入口
2 书店
3 教堂入口
4 教会唱诗班区域
5 祭坛
6 礼堂
7 哈普大道上的东正教中心入口
8 学院路上的东正教中心门厅
9 食堂
10 公寓
11 服务设施
12 教区中心
13 衣帽间
14 教堂
15 咖啡厅
16 展厅
17 教室
18 停车场
19 操场

de Poste
TAXIS
BOSQUET

勃朗峰地区私人木屋

这个面积为600平方米的私人木屋是对勃朗峰地区乡土建筑风格的一种重新诠释。这座建筑由钢筋混凝土和木制框架建造，其外表是由花纹染色橡木、葡萄牙石板和金属板构成的。这种葡萄牙石板每一块都是精挑细选出来的，以用于房屋的基础和屋顶部分。木屋的外部固定附着物都采用了煤黑色的金属材料。

“我们把从有30年历史的农舍上拆卸下来的木材用在了这座乡土感十足的新木屋中。这种转换恰到好处，是一种传统与现代的混搭。那么什么是现代木屋呢？找到这个问题的答案真是一个极好的挑战。”让-米歇尔·威尔莫特说。建筑师需要对建设项目进行重新诠释、对材料重新审视，在展现现代性的同时也要兼顾到过去的传统风格。这座房屋的设计就是要努力去探索哪些东西在山区建筑中是最值得注意的，同时也要力图避免在阿尔卑斯山建造豪华住宅而经常遇到的那些误区。

对于这个项目，威尔莫特工作室仔细研究了在阿尔卑斯山区木屋所使用的各种木制饰面，力图参考各种山区建筑，同时还要体现新木屋的现代性。朝南的阳台和游泳池融入了木屋的自身结构当中。温泉设有石墙和地板，与外面使用的石板相似。同样的内外连续性也体现在木屋的上层，那里采用了全高度的开口式格局。

项目地点 / 法国勃朗峰地区
完工时间 / 2016年
建筑面积 / 600平方米

位置平面图

西侧立面图

南侧立面图

北侧立面图

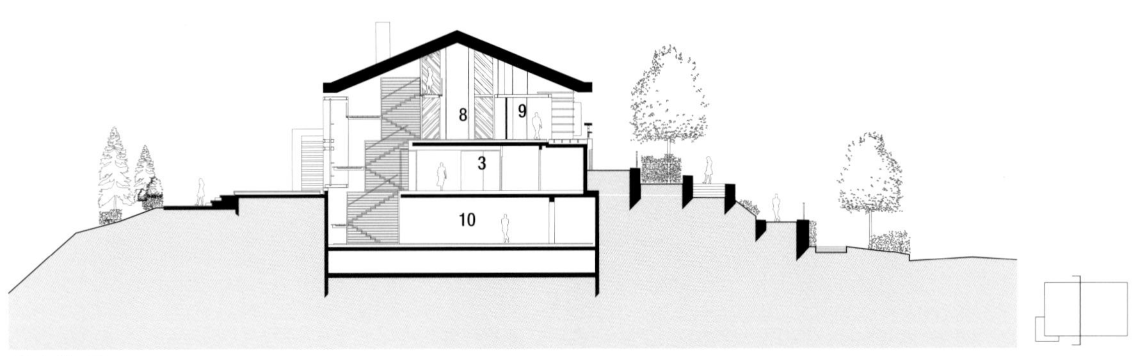

横向剖面图

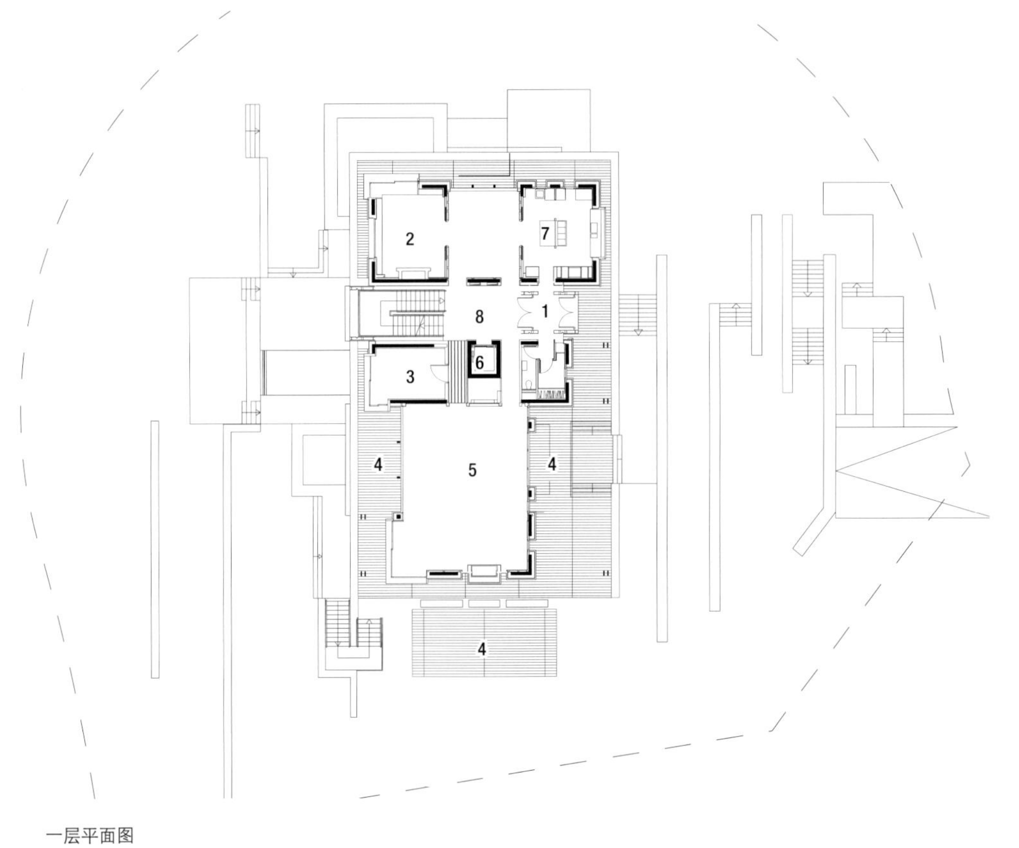

一层平面图

1 入口
2 小休息室
3 卧室
4 阳台
5 客厅
6 电梯
7 厨房
8 门廊
9 门厅
10 停车场

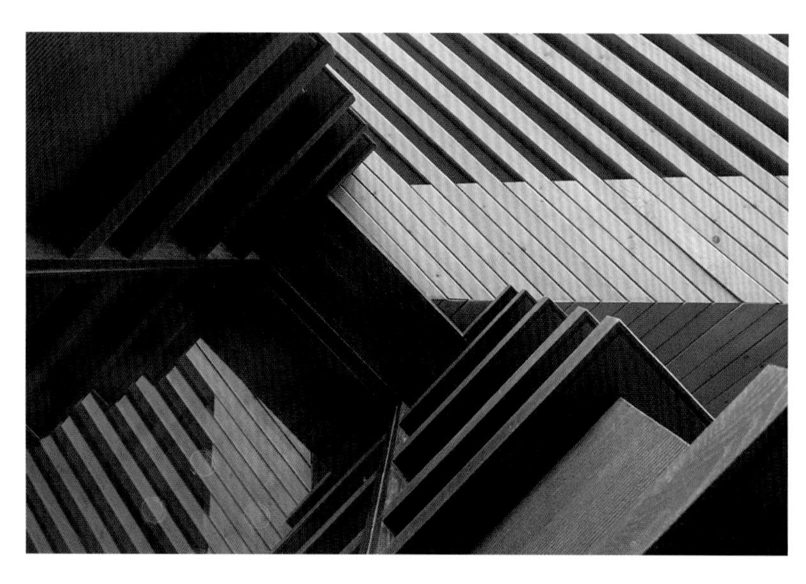

埃法日集团总部

项目地点 / 法国韦利济－维拉库布莱
完工时间 / 2015年
建筑面积 / 24 000平方米

建筑面积达24 000平方米的埃法日集团总部建有办公室、展厅、图书馆、公司自助餐厅和停车场等。这是一座外包铝钢架结构的混凝土建筑，外立面是通风良好的幕墙。该项目建在法国土木工程公司埃法日集团的旧址之上，这个公司是埃菲尔铁塔设计师古斯塔夫·埃菲尔创立的公司的后继者。新大楼由三个不同的建筑体组成，打算容纳1200名至1600名员工。

新建筑通过“桥”与原有建筑相连，组成了皮埃尔·贝尔热园区的一部分，它将整合公司的所有业务。为了表现集团总部的严肃性与功能性，这座建筑凸显了这家建筑工业公司在钢铁与混凝土这两方面的专长。三个建筑体中有两个的外立面由突出的棱柱状灰色铝合金边框窗户组成。高度抛光的铝合金窗框形成一种规则图案，使外立面给人以方正有序之感。主大厅内有一道雄伟的钢制螺旋楼梯通往下面的花园楼层。大厅是大楼的公共区域——自助餐厅和公司餐厅。这个楼梯是一件经过现场喷砂与涂漆的、真正的雕塑品，而一组日光灯管则模仿着螺旋上升的楼梯，一根根地悬挂其上。

明亮的大厅从北到南，面积达560平方米。它作为展览空间，用移动显示设备展示着埃法日集团的专业技术。这里的格局专门为大厅及临时展览而设计，可以举办大型模型展。这个大厅并不华丽，但反映了室内设计的功能性美学：浇筑混凝土、釉面砖地板、双层隔音天花板、黑色金属门与窗框等。从大厅里的每一处，访客们都可以看到正下方的花园，仿佛在大楼周围形成了一只天然的植物大碗，绿意盎然。

位置平面图

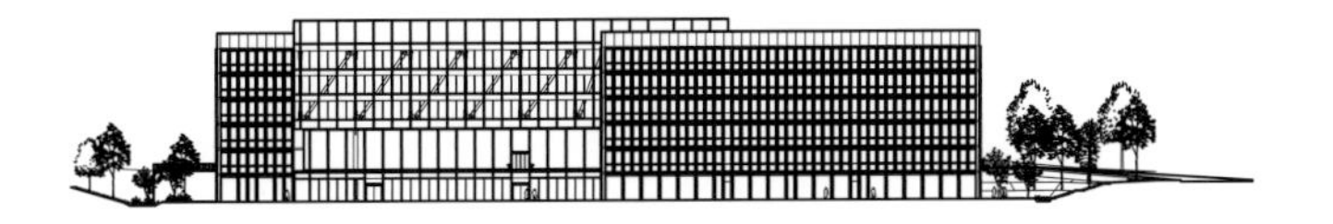

南侧立面图

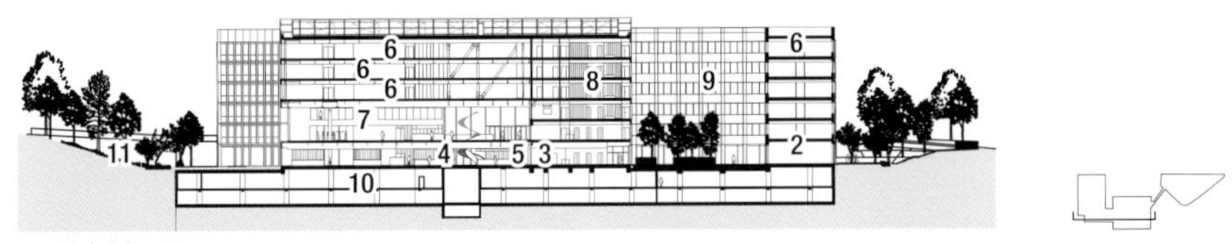

纵向剖面图

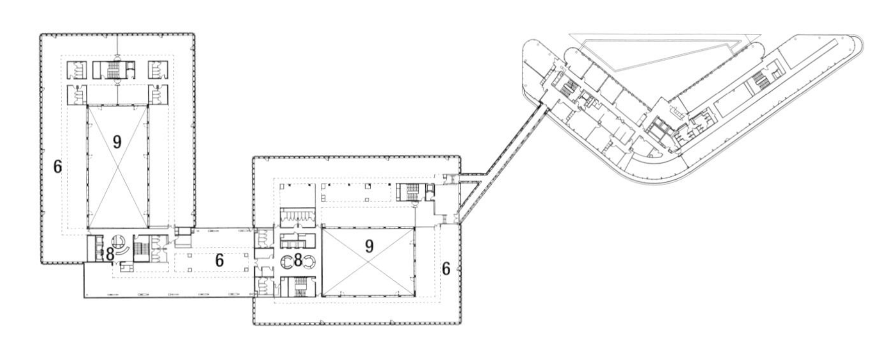

五层平面图

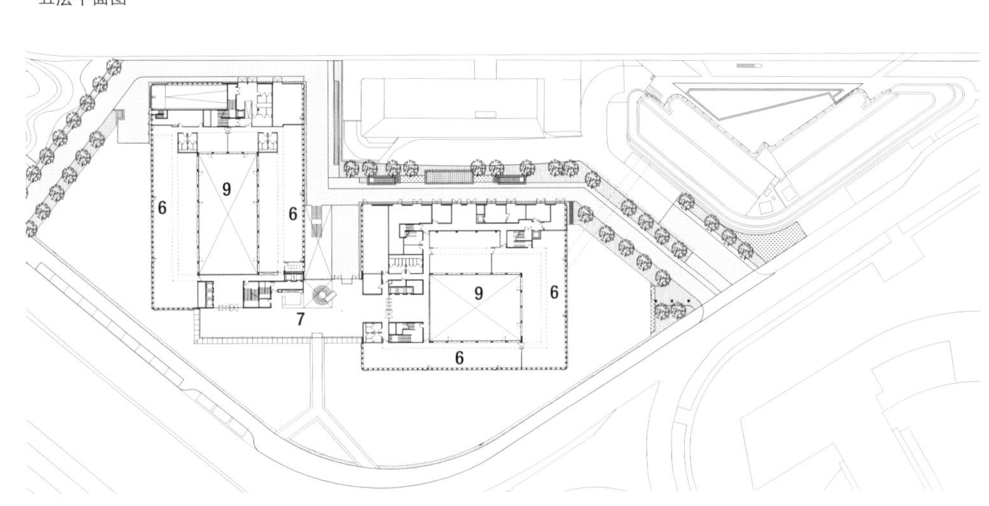

一层平面图

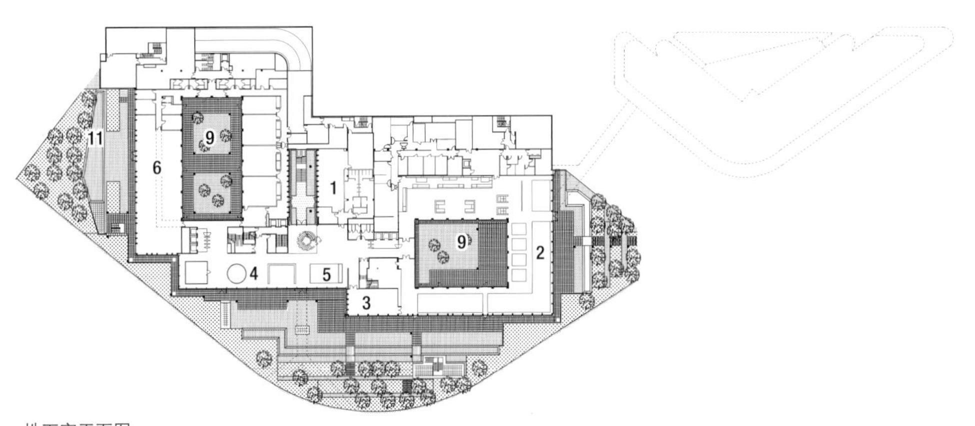

地下室平面图

1 健身室
2 餐厅
3 图书馆
4 娱乐室
5 咖啡厅
6 办公室
7 大厅
8 会客室
9 天井
10 停车场
11 花园

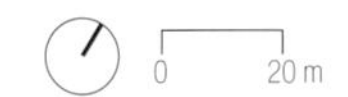

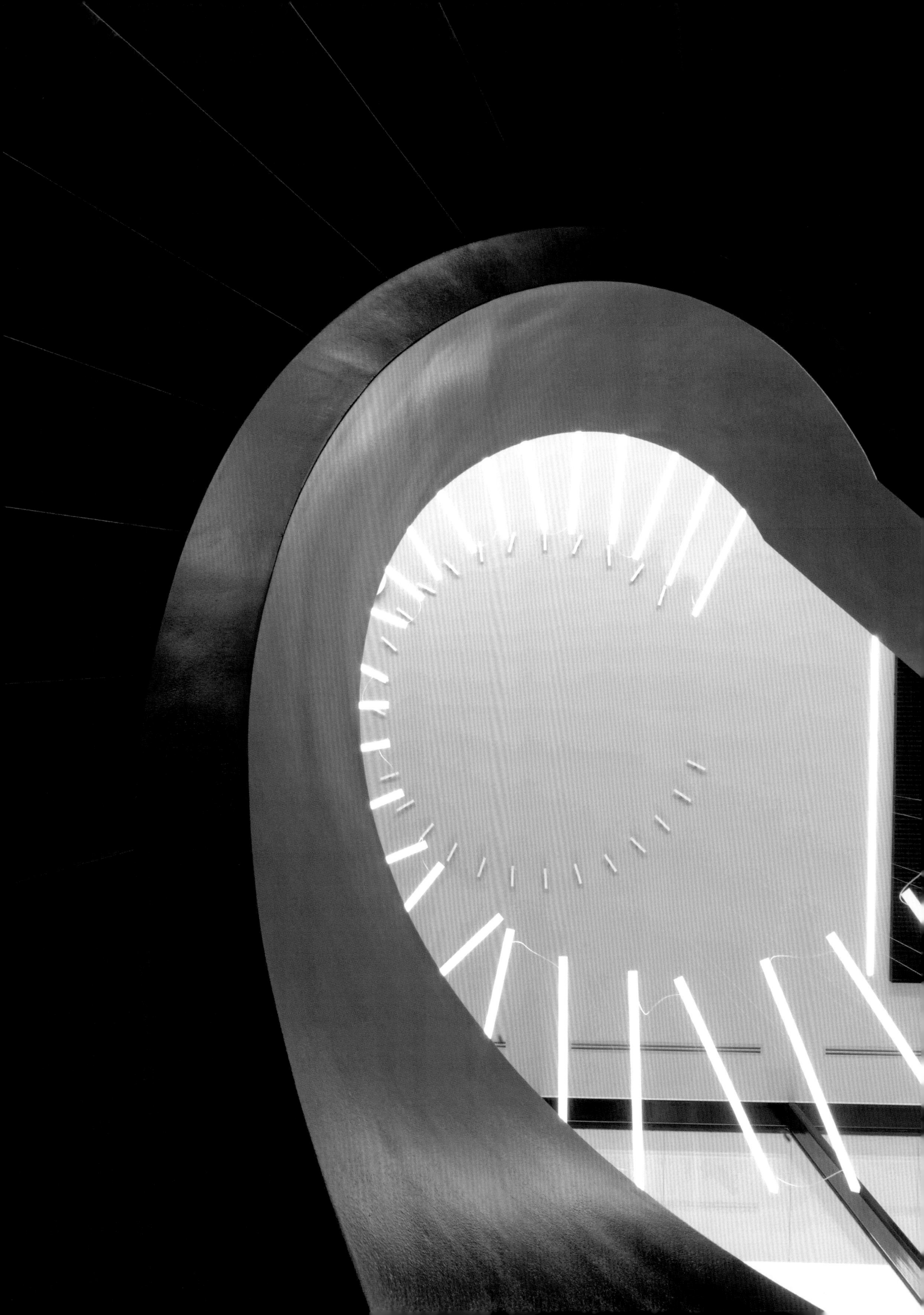

珀夏尔大厦

“哈伍德村”项目位于哈伍德的“17城”街区，是达拉斯上城区的新成员。该建设地点是夹在北面麦金农街、南面北哈伍德街和东面沃尔夫街之间的一个矩形区域。作为一个城市生活的新标杆，这个项目试图为达拉斯的城市扩张带来新的活力。

规划的项目包括一座31层的住宅——珀夏尔大厦、一座27层的办公大楼（名为“七号楼”）以及一个有商店和餐馆的景观广场。珀夏尔大厦达到113米的高度，并成为哈伍德区的一个“门户”式建筑。本项目增加超过4600平方米的零售与餐饮空间、超过27 800平方米的办公室空间以及230套公寓。珀夏尔大厦的公寓无论是单居室还是两居室，都有一个大型私人阳台以满足人们现代生活的需要。

珀夏尔大厦预示着这座城市的扩张并没有止步，而是向外延伸，它使这里充满了活力。这座有着雕塑般造型和曲线的大厦受到了许多自然法则方面的影响，但最重要的推动力还是建设地点的环境本身。带磨砂玻璃栏杆的曲线式阳台为这座高楼提供了一种柔和的外观表现，与相邻的直线式建筑相呼应。为了避免大厦对邻近建筑形成过大的压迫性，在与之相对的一面，大厦的外形又从曲线形状逐步恢复到了常见的笔直模样——这也是一种可以让它能够更好地抵御风力的造型。

“七号楼”基于工作场所、公共区域、气候和现有条件等，从内到外都被设计成了一个远望这座城市的窗口。整个办公大楼除了最底下的三层以外都是整齐划一的玻璃窗，而这三层在麦金农街上则转变成一种挑战地心引力的悬臂式结构。由于这里靠近美国航线中心体育馆和艺术区，所以哈伍德村项目既满足了附近街区的需求，同时也在城市化进程中为社区提供了一个标志性建筑。

项目地点 / 美国德克萨斯州达拉斯
完工时间 / 2017年
建筑面积 / 42 800平方米

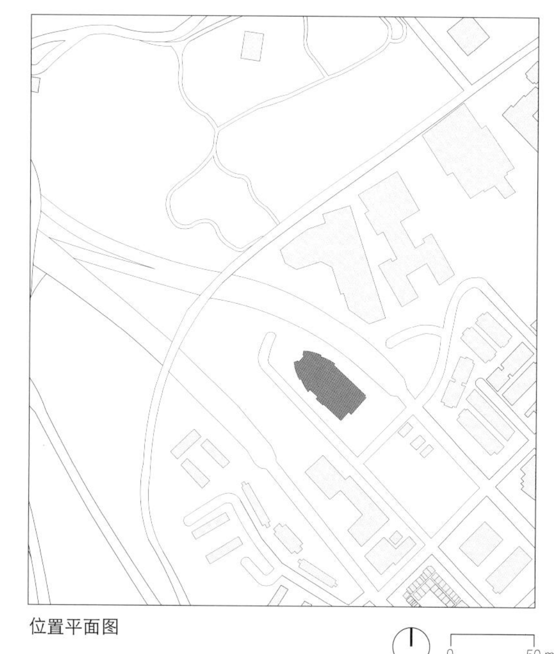

位置平面图

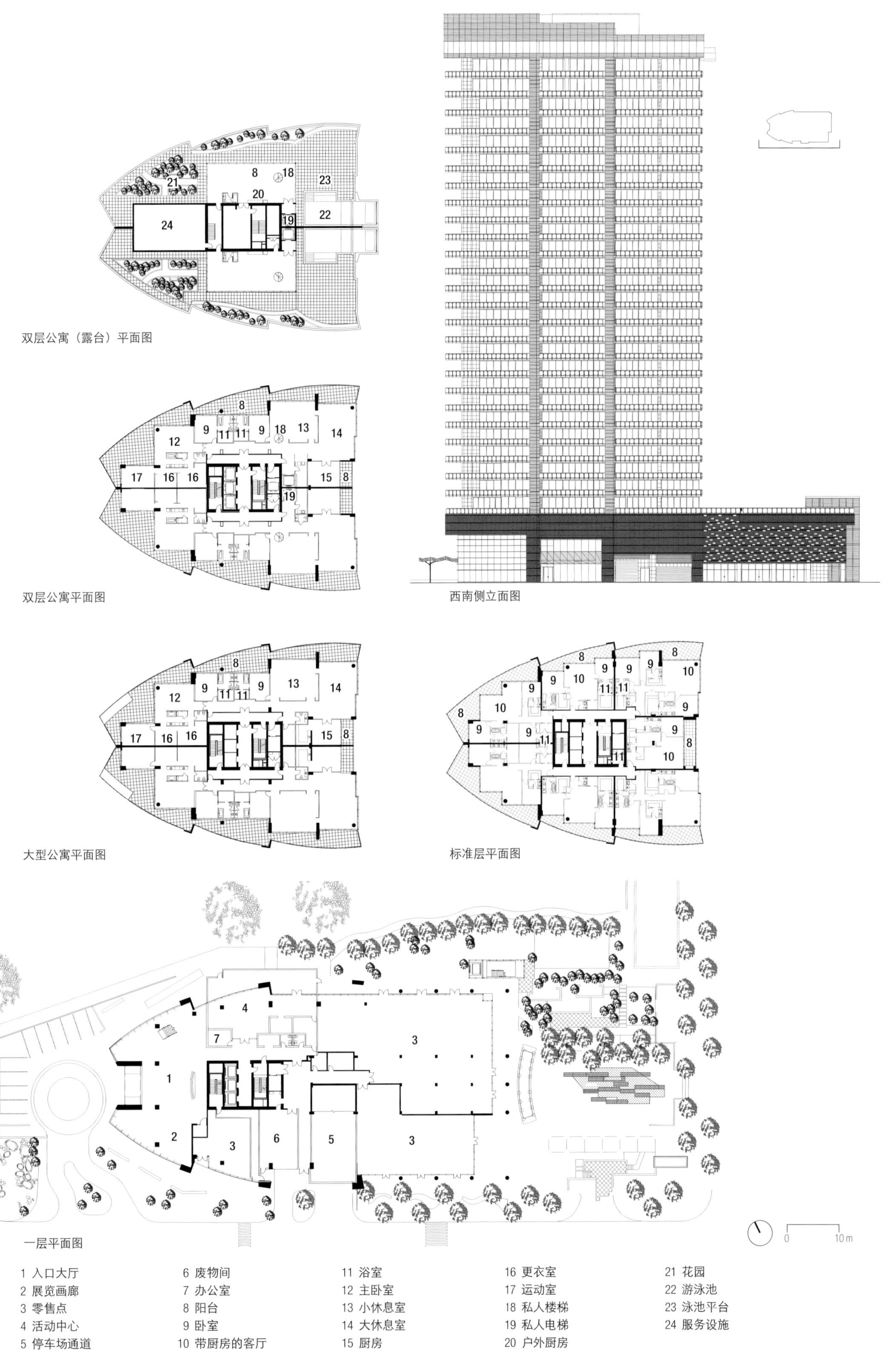

双层公寓（露台）平面图

双层公寓平面图

西南侧立面图

大型公寓平面图

标准层平面图

一层平面图

1 入口大厅
2 展览画廊
3 零售点
4 活动中心
5 停车场通道
6 废物间
7 办公室
8 阳台
9 卧室
10 带厨房的客厅
11 浴室
12 主卧室
13 小休息室
14 大休息室
15 厨房
16 更衣室
17 运动室
18 私人楼梯
19 私人电梯
20 户外厨房
21 花园
22 游泳池
23 泳池平台
24 服务设施

紫罗兰大厦

紫罗兰大厦建在摩纳哥东侧的圣罗马区，建成后将是一座22层、130米高的住宅大厦。它位于紫罗兰街和特内奥大道之间，拥有73间豪华公寓。这里将有一个家庭综合服务中心、一个水疗中心、零售空间以及三个游泳池，其中有两个泳池建在私人公寓的屋顶上。

这座包含6个地下停车楼层的钢筋混凝土建筑经过精心设计，将最大限度地获得南边地中海的视野。不锈钢、石材和玻璃用于建筑的外包层，其尽可能减少能源消耗的设计让人眼前一亮。可移动的遮阳板和广阔的阳台可以为人们提供有效的遮蔽。

项目建成后，高效能的紫罗兰大厦无疑将获得英国建筑研究院绿色建筑评估体系优秀级环保认证。建筑的用水、加热及冷却等区域都经过精心设计、控制，以减少大厦的碳排放量。让–莫斯负责项目的景观设计，他的工作室在不远处的卡布里斯，而室内装饰则由欧若・伊图负责。这座大楼给人的感觉是在地面附近从花园与阳台中央拔地而起，留下了它身下周围的绿色空间。这些绿色植物都种植在建筑的围墙之内。由于这座建筑只占据了建设地点28%的面积，所以留下了充足的绿地空间。该项目的客户是米歇尔牧师团。

项目地点 / 摩纳哥
完工时间 / 2020年
建筑面积 / 21 000平方米

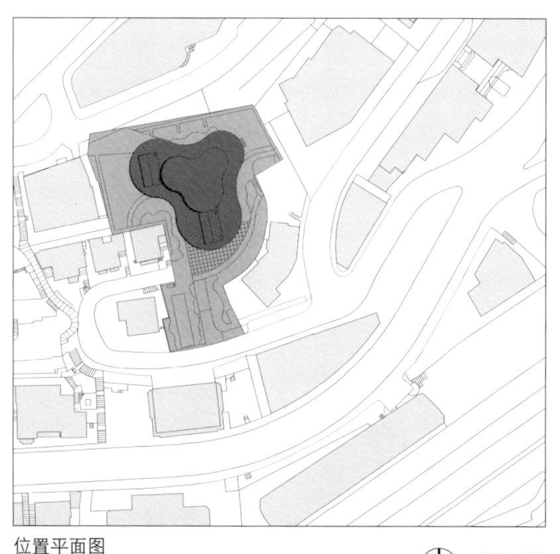

位置平面图

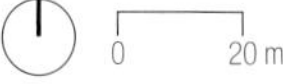

南侧立面图

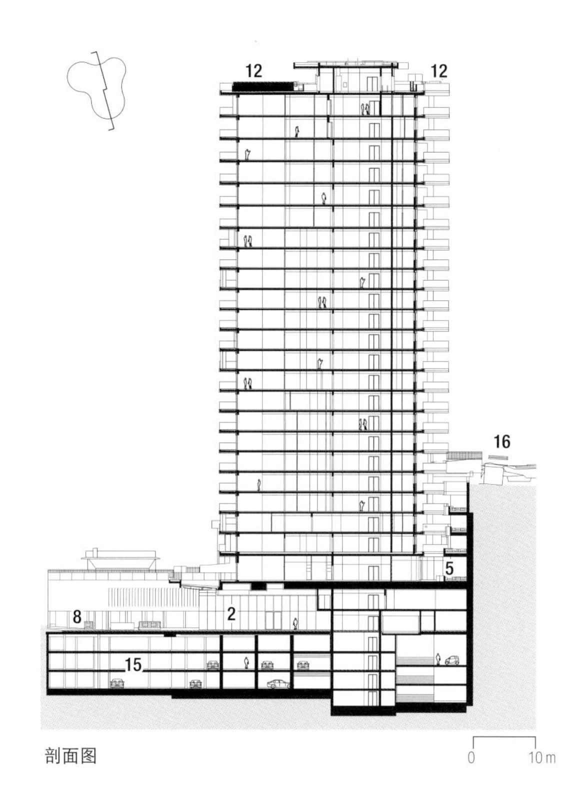

剖面图

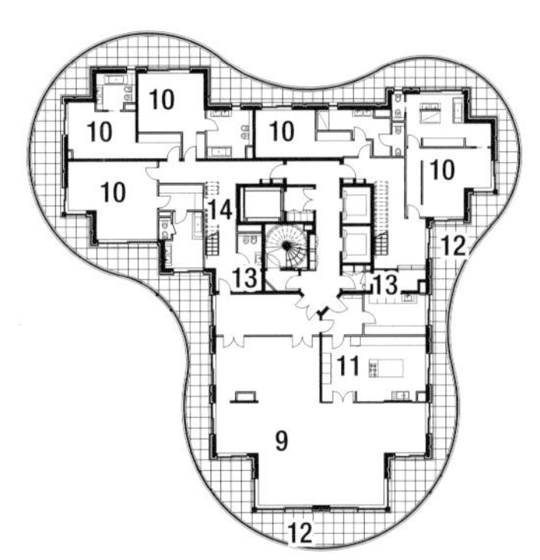

三室公寓平面图 (22层)

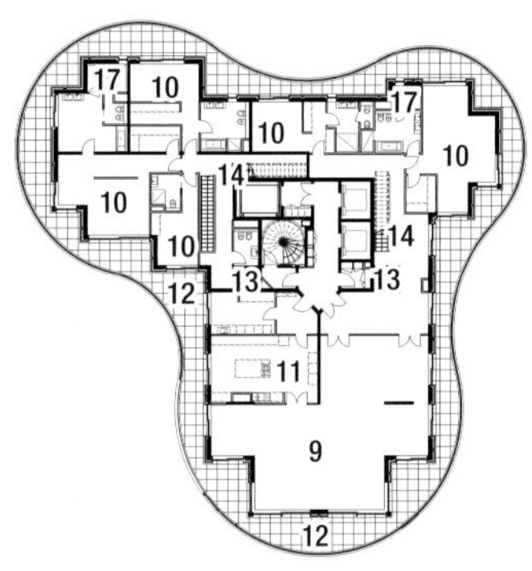

三室公寓平面图 (23层)

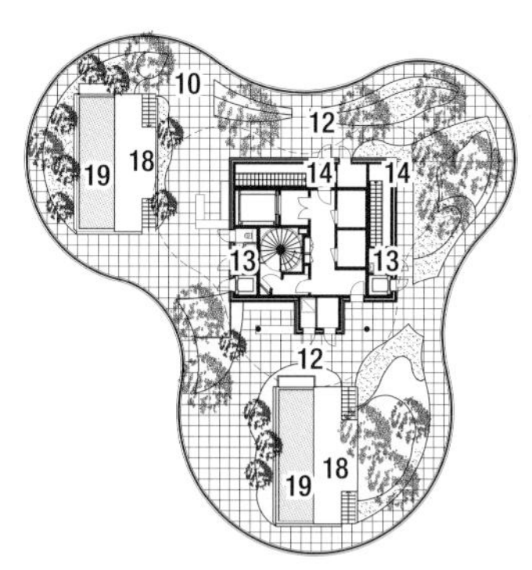

三室公寓平面图 (24层)

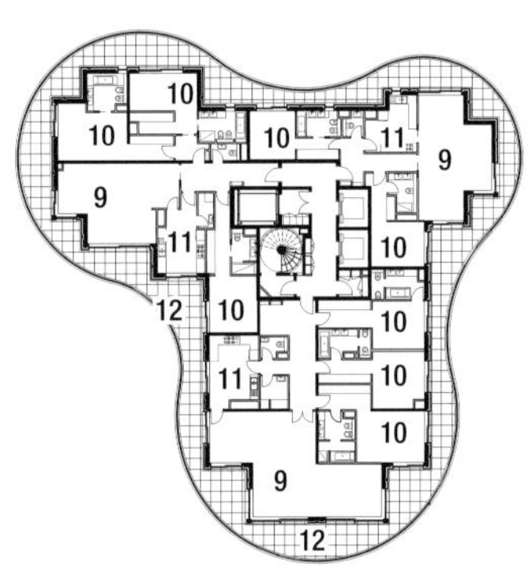

标准层平面图

1 接待大厅
2 零售点
3 安全控制室
4 自行车停放处
5 摩托车停放处车
6 废物间
7 服务设施
8 停车场通道
9 客厅
10 卧室
11 厨房
12 阳台
13 私人电梯
14 私人楼梯
15 停车场
16 第6层的北通道
17 桑拿
18 甲板
19 游泳池

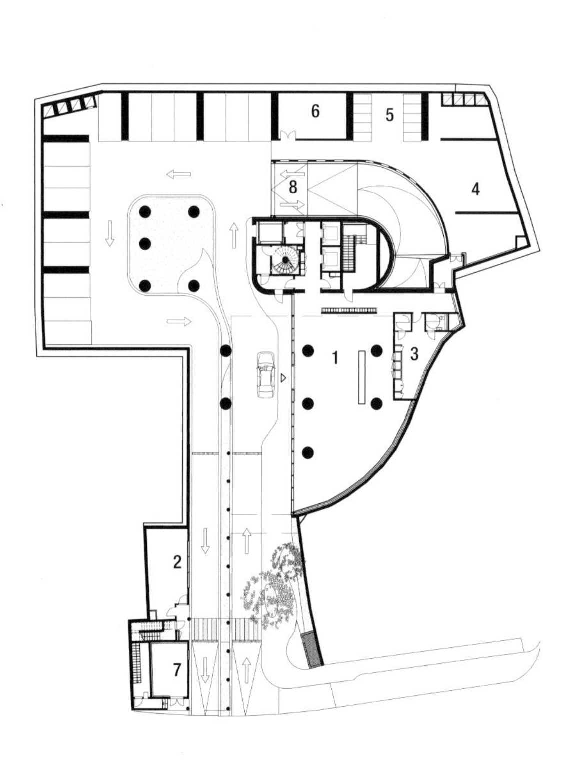

低层入口平面图

RENOVATION PROJECTS
翻新项目

瑞瑟夫别墅酒店

让–米歇尔·威尔莫特被要求对这座20世纪70年代的建筑进行重新设计，使其成为一座带有1000平方米水疗中心的五星级酒店。该项目的总建筑面积为8300平方米，场地面积9000平方米。这个酒店的海滨部分还有12套出租别墅。酒店的7间客房与16间套房面积大小从50到100平方米不等，每个都设有私人阳台或花园。

这座钢筋混凝土建筑主体采用光滑的水泥立面，外表采用干砌石，地板使用天然石材，外面部分则使用了木质铺面板。这座酒店位于种满了圣栎与石松的山上，可以欣赏到地中海的美景——博坡图海湾和泰尔莱特角。原来建筑的灰泥砖砌外立面被修复如新。室内的无边泳池朝向大海，好像与之融为一体。建筑师将别墅酒店前面的地势降低，使阳台上举办的各种活动不会干扰游泳者的视野。“这里的植被与海景就是我画作的颜料。”威尔莫特说。这个项目赢得了2011年《壁纸》杂志设计大奖的“年度最佳新建酒店”奖。

项目地点 / 法国拉马蒂埃尔
完工时间 / 2009年
建筑面积 / 8300平方米

位置平面图

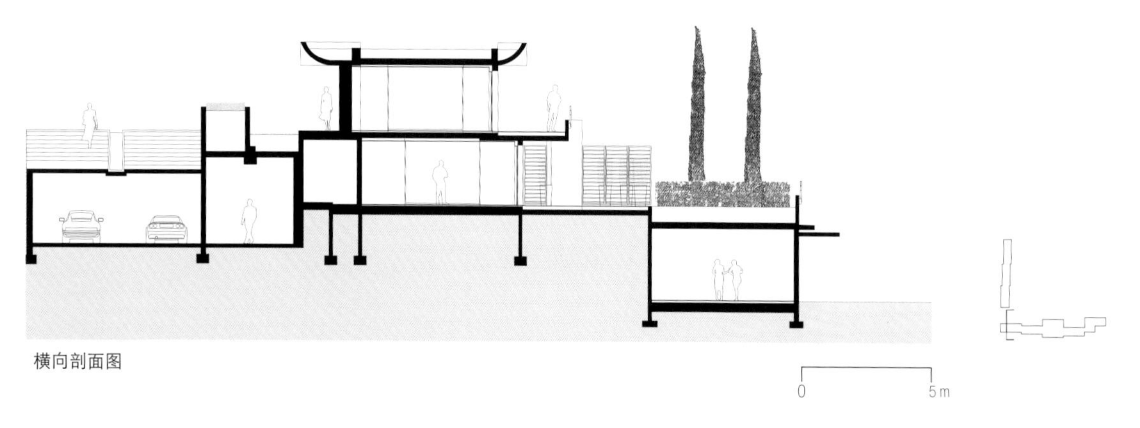

横向剖面图

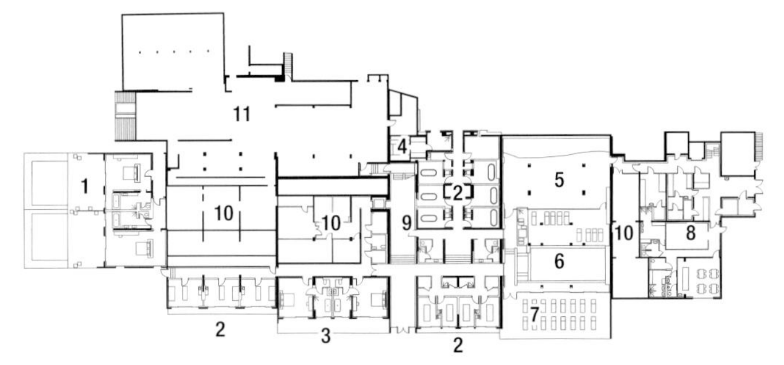

三层平面图

二层平面图

一层平面图

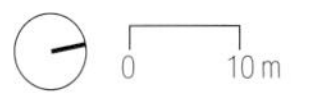

1 带花园的豪华客房
2 康乐房
3 豪华客房
4 土耳其浴室
5 健身房
6 室内泳池
7 泳池阳台
8 服务设施
9 卫浴设施
10 设备间
11 停车位
12 配有双人床和两张单人床的卡马拉特套房
13 带花园的小型豪华套房
14 带花园的尊享套房
15 带花园的豪华套房
16 带花园的高级客房
17 带花园的豪华客房
18 员工房
19 入口
20 休息室
21 饭厅
22 餐厅的阳台
23 厨房
24 尊享套房
25 带私人阳台的泰尔莱特套房
26 游泳池
27 泳池房

让-皮埃尔·雷诺工作室

项目地点 / 法国巴比松
完工时间 / 2010年
建筑面积 / 240平方米

这所房子的主人是与建筑师让-米歇尔·威尔莫特在设计工作上密切合作的著名法国艺术家让-皮埃尔·雷诺。雷诺在离巴黎不远的巴比松镇购买了这座诺曼风格的名为霍顿斯葡萄园的房子。这里以巴比松画派闻名于世，其中包括让-弗朗索瓦·米勒和泰奥多尔·卢梭等19世纪中叶的重要画家为代表。这种艺术参照对于雷诺而言不可能视而不见，即便他自己的作品中没有体现出太多，但还是与巴比松画派的写实主义绘画有着莫大的关系。

建筑师拆除了这所20世纪60年代老房子的后期增建部分，并增加了一个大型的玻璃内部空间，建造了475平方米的阳台。这座拥有百年历史的小型北欧式亭屋（63平方米）坐落于此，它还曾被作为1939年世界博览会的一部分。威尔莫特应艺术家让-皮埃尔·雷诺的要求将之全部涂成了黑色，用来展示雷诺的作品。尽管让-皮埃尔·雷诺将这所房子称为他的工作室，但他仍不会在传统的工作室环境中创作作品，而是主要使用一栋240平方米的房子（在拉加雷讷科伦布）用于展示他已经完成的作品，如他的代表作——镀金的巨型花盆是一件坐落在观赏性花池中的大型艺术作品，从拉加雷讷科伦布的工作室空间一直延伸到了外面。

这个项目的景观设计由奈维欧·卢耶完成。这座房产挂牌出售后，让-皮埃尔·雷诺和他的妻子达芙妮就出手购得了此处。雷诺向来是建造（或拆除）他自己的房子，将之作为他艺术创作过程的一部分，他在巴黎附近的塞勒-圣克卢和加雷讷科伦布的两个工作室就已经证明了这一点。

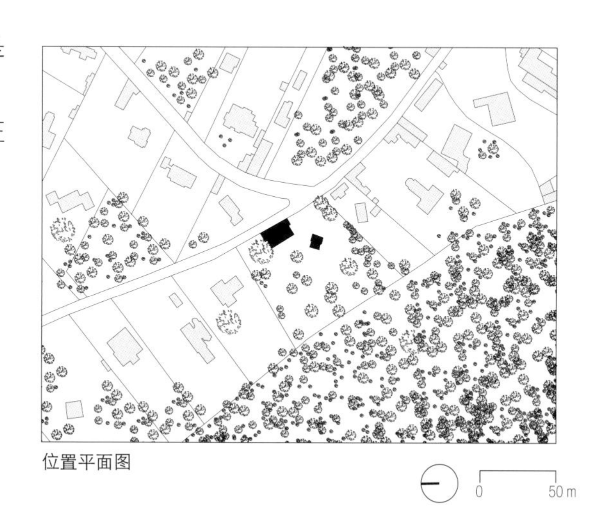

位置平面图

0　50 m

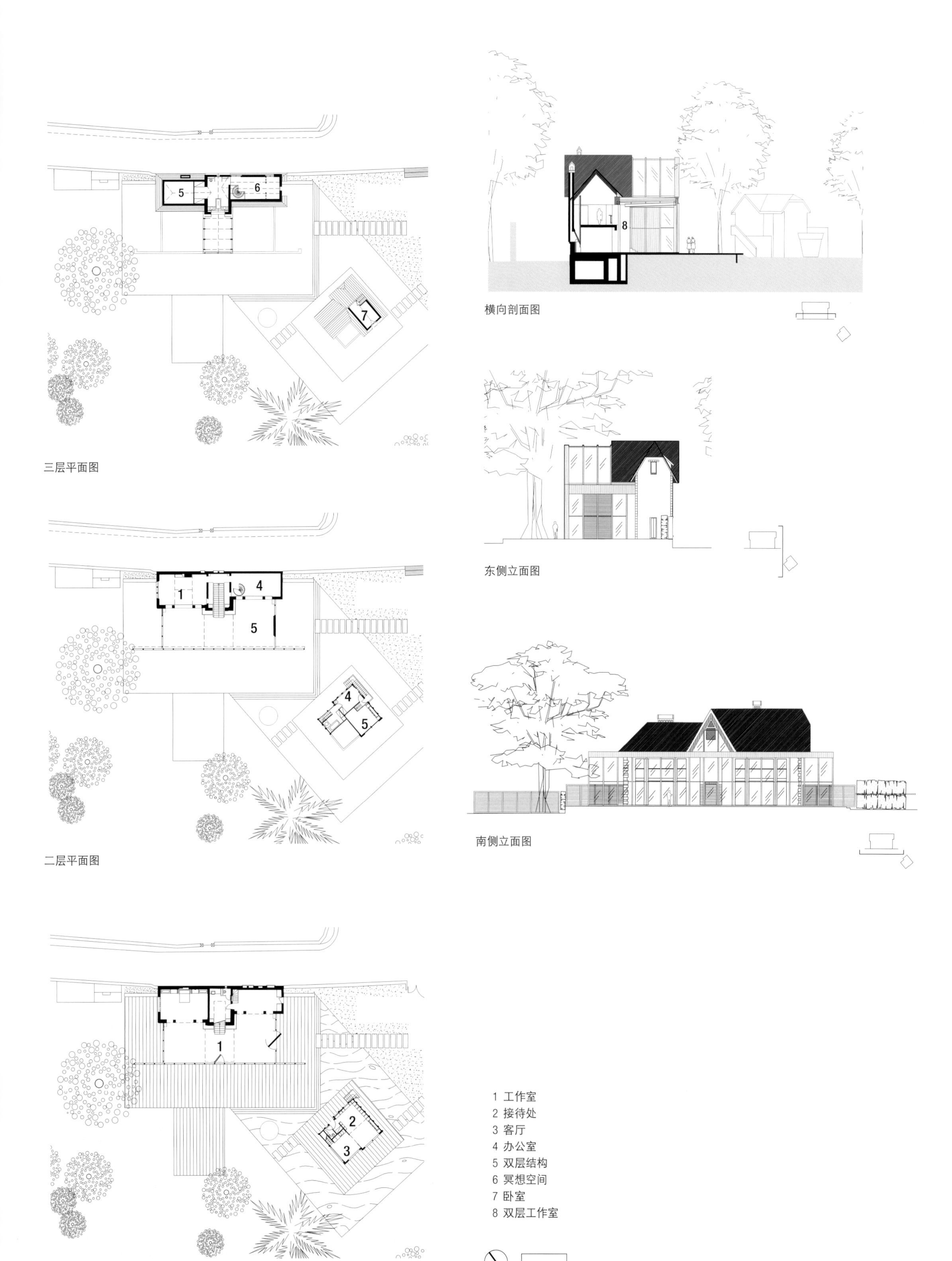
5
6
7
三层平面图
1
4
5
4
5
二层平面图
1
2
3
一层平面图
8
横向剖面图
东侧立面图
南侧立面图
1 工作室
2 接待处
3 客厅
4 办公室
5 双层结构
6 冥想空间
7 卧室
8 双层工作室
0
5m

文华东方酒店

应文华东方集团的邀请，让–米歇尔·威尔莫特承担了现有办公楼的重组工作，并将其改造成为一家拥有99间客房和32间套房的五星级酒店。它也是在法国第一个获得法国绿色建筑标准认证（HQE）的豪华酒店。项目所用的主要材料是混凝土，而天然石材则用于临街立面，天然石材和石灰泥构成的外立面俯瞰着下面的花园。项目的景观设计是与建筑师经常合作的景观设计机构奈维欧–卢耶。

这座文华东方酒店位于圣奥诺雷路和塔波山路之间的一个面积为3000平方米的地块上。它由三栋建筑物和两个内部庭院组成，这个建筑群在圣奥诺雷路上只有一个临街立面。建筑师决定拆除两个庭院之间的建筑物，这样可以给未来的酒店留出一个宽阔的中央花园（占地面积21米×22米）。而双层楼高度的酒店水疗中心则位于中央庭院的下方。建筑的四个外立面基于颇有历史纪念意义的正门与临街立面上的窗户进行了彻底的改造，对于新酒店来说做出了新的诠释。酒店室内装饰工作由著名女设计师西博尔·马格里完成。这座五星级的文华东方酒店已经获得了“皇宫级豪华酒店”评级（比五星级酒店标准更高）。

项目地点 / 法国，巴黎
完工时间 / 2011年
建筑面积 / 22 000平方米

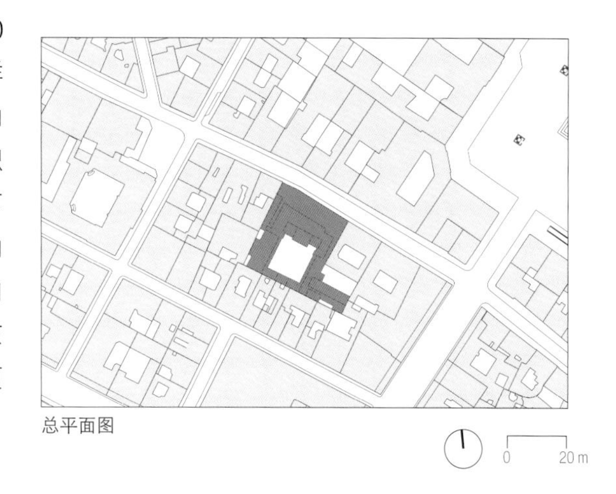

总平面图

街道一侧立面图

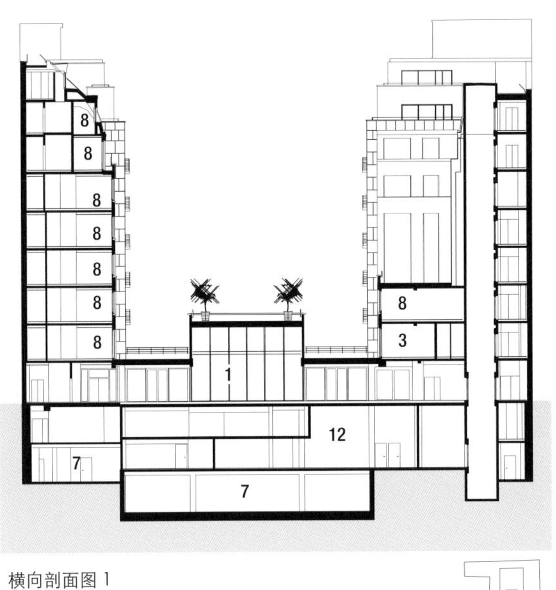

横向剖面图 1

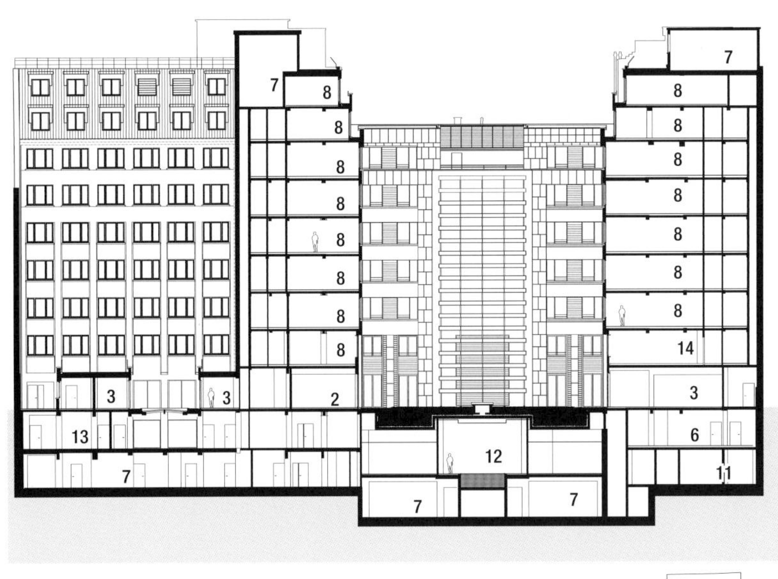

横向剖面图 2

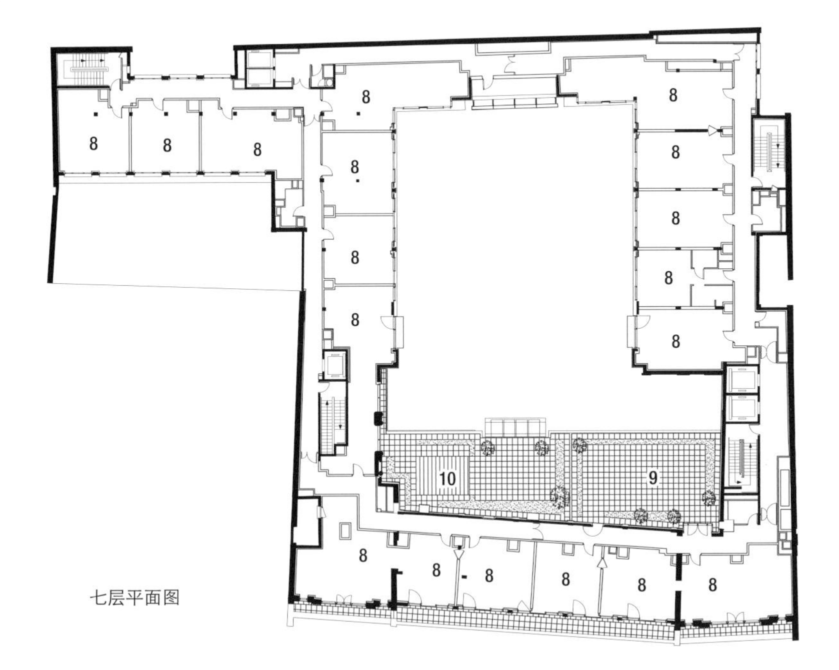

七层平面图

1 接待大厅
2 酒吧
3 餐厅
4 商店
5 花园
6 厨房
7 服务设施
8 卧室或套房
9 阳台
10 私人阳台
11 大厅
12 水疗中心
13 衣帽间
14 会议室

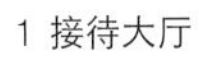

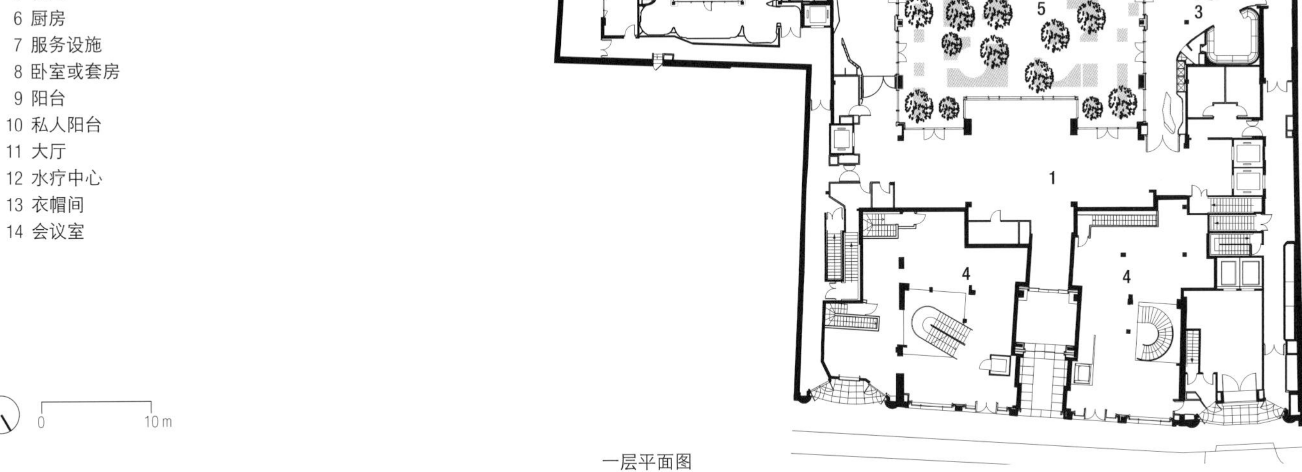

一层平面图

DSQUARED2
DSQUARED2
DSQUARED2
opening soon

ORIENTAL
PORTS
1961
COMING FALL
PORTS
1961
COMING FALL
PORTS
PORTS
PORTS
1961
COMING FALL

百德诗歌酒庄

项目地点 / 法国波亚克
完工时间 / 2015年
建筑面积 / 7700平方米

波亚克地区的第五家葡萄酒厂百德诗歌酒庄获得了列级酒庄排名（波尔多葡萄酒分级体系，1855年建立），它于2009年开始改建工作，并要求威尔莫特建筑事务所设计新的技术性设施，尤其是一种由重力驱动的酿酒设施。事实上，这里的葡萄酒已经不再被泵送到发酵罐之中，而是被直接传送到一种可升降的不锈钢大桶中，这种酒桶可以从地下室一直升到建筑物的二楼。这种设备也通过采用戏剧性的打光营造出了一种表演性的展示效果。

改造工作包括对酒庄的全面整修。原有的建筑物被拆除，由一套中央设备取代了原来酒桶间、酒窖和装瓶间的长方体建筑组合（建筑面积达1860平方米）。与此同时，这座酒庄通过玻璃结构的横向扩展而进行了扩大，这个玻璃建筑现在作为品酒室和商务洽谈室。在品酒室可以尽享南面葡萄园的美景，而里面的家具则是由维多利亚・威尔莫特特别设计的。透明度是这两座建筑物共同的主题。酒桶间的酒桶在建筑物底部的玻璃墙后面被戏剧性的排列和打光，使得酒桶间的褐色外立面如同直接被116只不锈钢酒桶托起来一样。“美感、比例、精致与舒适，这些会让人不由自主地产生幸福感，我就是那些捍卫并促进这种理念的人之一。”让–米歇尔・威尔莫特说。这座桶仓的地下室是用白色钢筋混凝土建成的，而在一楼和二楼的酒桶间则采用一种古铜色金属结构。

百德诗歌酒庄位于波尔多附近的上梅多克地区，而整个波亚克产区包括如拉图尔酒庄、拉菲・罗斯柴尔德酒庄、木桐酒庄等著名葡萄酒品牌。威尔莫特设计的酒庄面积达7700平方米，使用了石英黑混凝土，酒桶间立面采用古铜色电镀铝板条，正面是大尺寸玻璃面板，外面的地面采用白色混凝土板，而里面则采用了美国胡桃木地板。

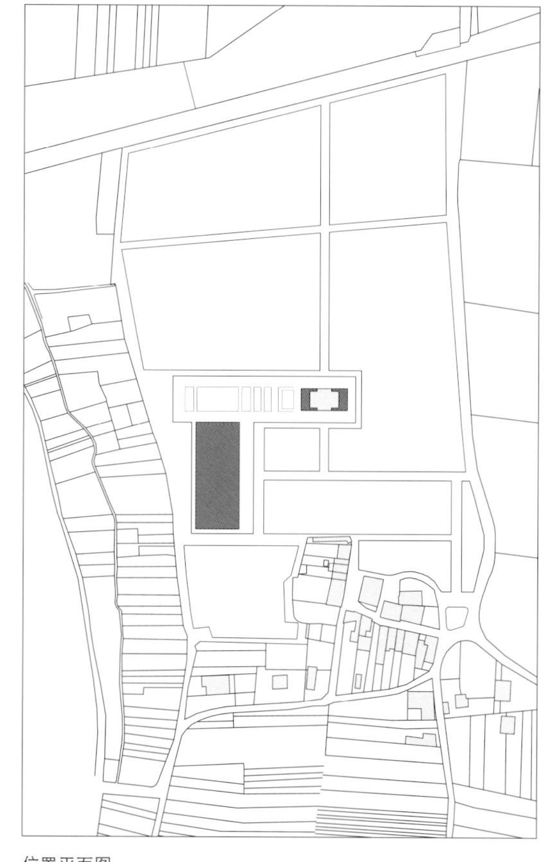

位置平面图

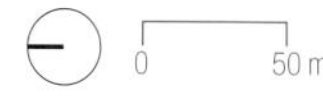

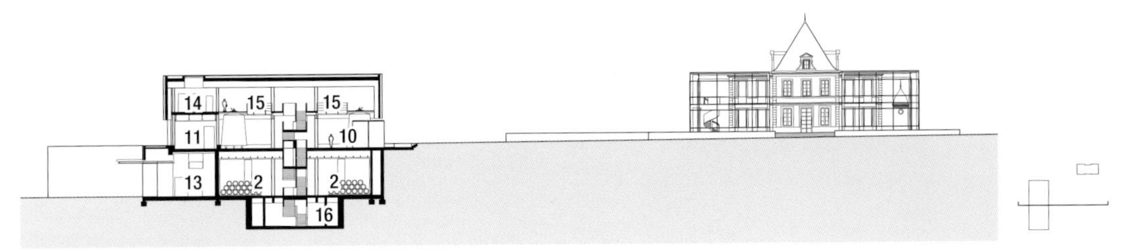

酒庄横向剖面图和西侧立面图

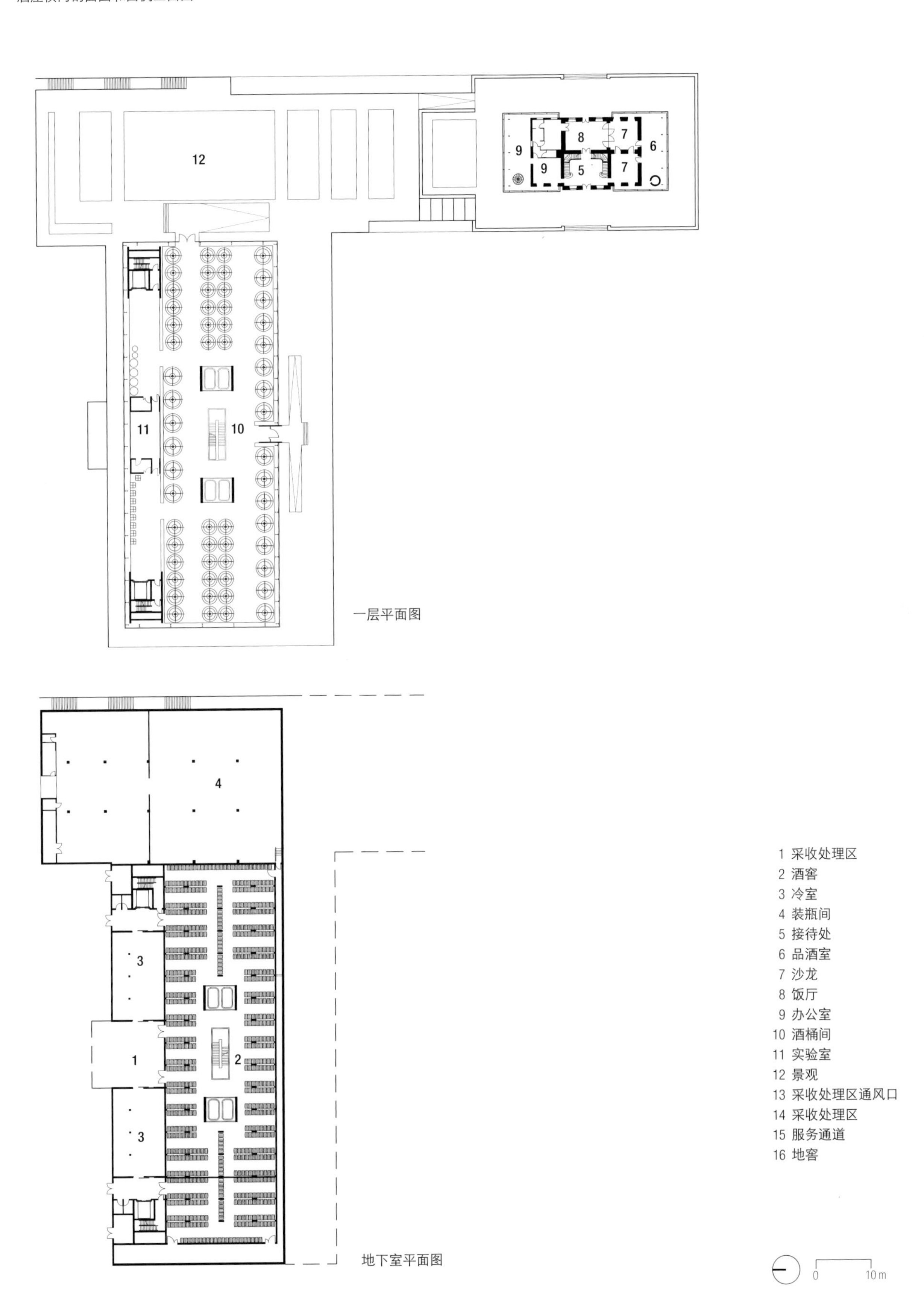

一层平面图

地下室平面图

1 采收处理区
2 酒窖
3 冷室
4 装瓶间
5 接待处
6 品酒室
7 沙龙
8 饭厅
9 办公室
10 酒桶间
11 实验室
12 景观
13 采收处理区通风口
14 采收处理区
15 服务通道
16 地窖

0 10m

德高集团英国总部

项目地点 / 英国伦敦
完工时间 / 2016年
建筑面积 / 6500平方米

全球著名户外广告与街道设施公司法国德高集团在英国的新总部占地面积6500平方米，位于伦敦市中心的帕丁顿地区，靠近埃奇韦尔路与帕丁顿盆地，建在一个往南是贝斯沃特与帕丁顿保护区，往北是帕丁顿开发区之间的地块上。该建设地点原有两栋独立的办公大楼，分别建于1968年和1970年，还有一个酒吧以及一些住宅公寓和零售空间等。

威尔莫特建筑事务所的方案是完全翻新这两座建筑物以使其达到最高规格，遵循英国文化协会办事处的指导，将接待区延长到原来由酒吧占据的地块中，扩建两座建筑物的七楼，使之连通并将这两座建筑的内部布置统一。通过各种材料的适当使用，该建筑形成了一种到帕丁顿保护区的适当过渡效果，使得普雷德街沿途的景观视野得到了显著的改善。

除了面向保护区方向的住宅外墙以外，这些建筑物的原有结构被彻底剥离。为了获得更大的接待区，大楼需要进行结构性修改：拆除原来的六根混凝土柱，用大型桁架代之；原来的主立面采用一种幕墙系统进行了改建，上面的电镀铝翅片结构突出了普雷德街的笔直特点。

一楼以黑色的印度花岗岩进行了改建，营造出了零售空间和接待区的墩座墙。由于原建筑的天花板较低，所以服务设施的大部分隐藏在一个横跨整个办公空间的中央轴线上。一楼的顶板底面和裸露的管道系统被涂成了黑色，使其“消失不见”，以给人一种天花板似乎更高一些的印象。同时，为了减轻天花板高度较低的潜在问题，还安装了吸音板，精心设计了内部照明。

该项目的设计旨在符合英国建筑研究院绿色建筑评估体系BREEAM 2011优秀级标准。这个包括了扩建、翻新、室内设计与墙面整修等工作的项目还涉及对原混凝土结构的拆除与采用钢材进行改建的工作，而项目中明显可见的材料则是混凝土、铝材和玻璃等。

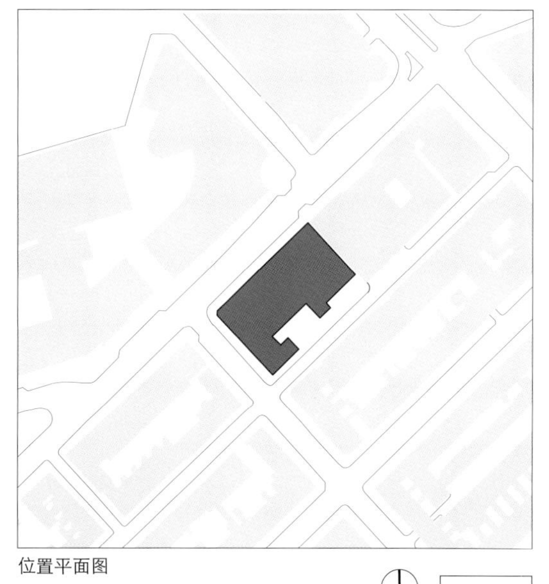

位置平面图

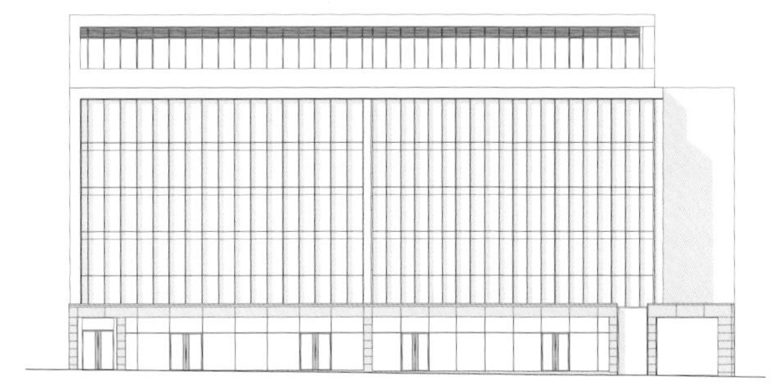

北向立面图

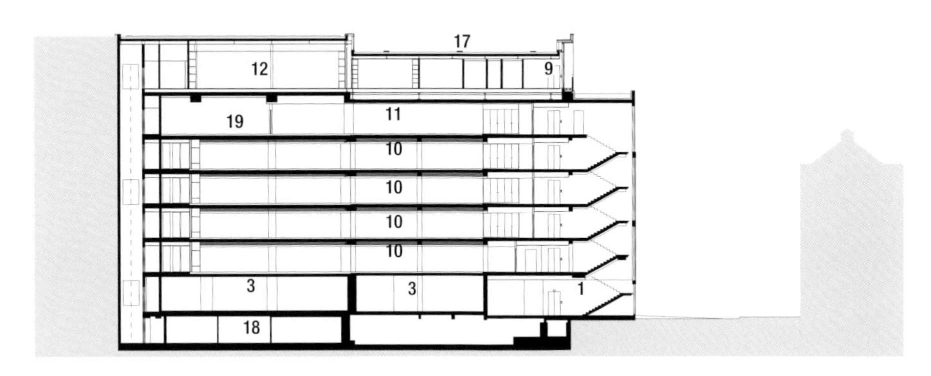

横向剖面图

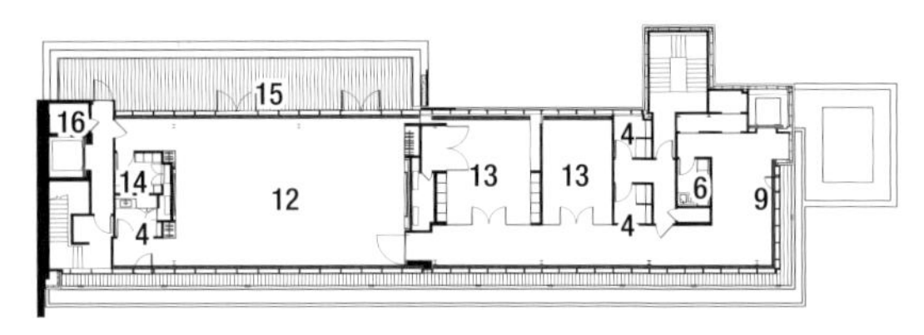

七层平面图

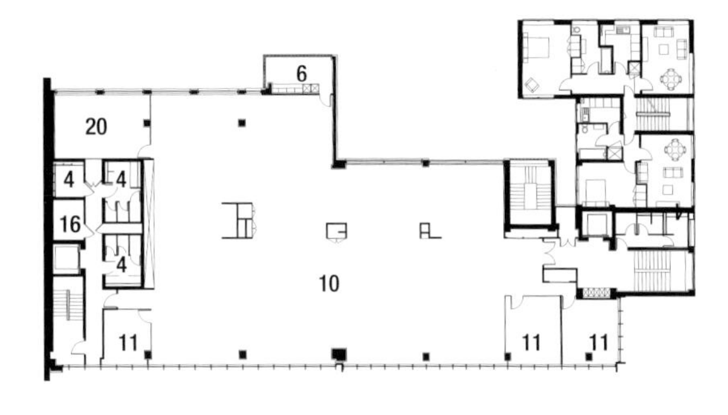

三层平面图

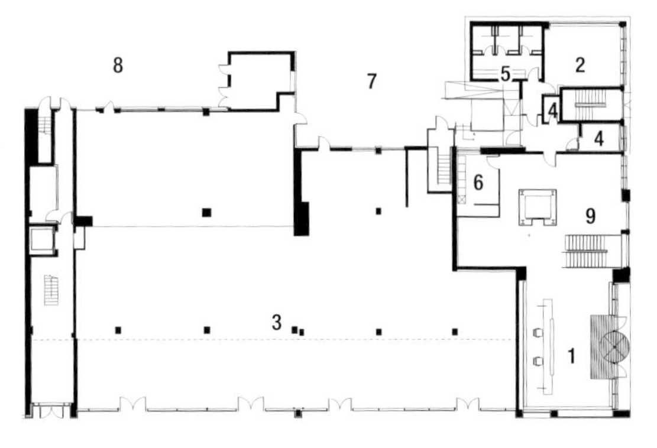

一层平面图

1 接待处
2 健身房
3 零售点
4 洗手间
5 淋浴间
6 小厨房
7 停车场（主管专用）
8 停车场（雇员专用）
9 休息室
10 办公室
11 会议室
12 总裁办公室
13 私人助理（PA）办公室
14 图书馆
15 阳台
16 仓储间
17 绿色植物
18 档案室
19 咖啡厅
20 主任办公室

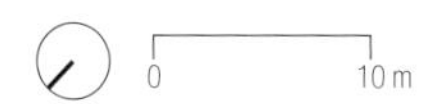

JCDecaux

F站初创企业园区

项目地点 / 法国巴黎
完工时间 / 2017年
建筑面积 / 34 000平方米

泽维尔·尼尔是一位著名的法国实业家，他拥有自由电信电话网络和《世界报》等项目，同时也投资其他领域的项目。他与让-米歇尔·威尔莫特一起合作，创建了这个园区以帮助那些有前途的创业公司。他们接管了一座由著名结构工程师尤金·弗雷西内于1927年建造的有盖大厅建筑，它具有非常轻的承重结构，于2012年被列为历史古迹。该建筑长度为304米，宽度在61到72米之间。

这座大厅曾经长期作为附近巴黎奥斯特里茨火车站的换乘点，如今被威尔莫特利用起来，改造后以反映年轻的数码创业者的生活方式与商业活动。F站初创企业园区现在拥有了共享的多用途空间以及围绕在各种服务设施周围的“村落”。有盖大厅的东南端设有一间与众不同的餐厅，全天24小时开放。这里保留了旧的装卸平台和钢轨底座，甚至还有几节餐车，使人能够回想起过去的车站。人们可以在繁忙而令人惊喜的餐厅通过花园的阳台眺望城市的风景。该项目为巴黎右岸的产业革命发展增添了新的活力。

有盖大厅视野开阔，它的中央桥跨部分是整个园区的主轴。其四周是各种公共的多用途设施，而且从这里的地面有入口可以进入地下的礼堂。这个主要区域有两个有盖通道穿过，它们旨在用作数码科技的窗口展示，使得本地居民可以亲近这个富有创造性的地方，同时把在这里的发现宣传给街坊四邻，以鼓励数码科技的创新活动。目前这两个区域是被铁路线相互隔绝的，所以除了鼓励大家讨论和交流之外，这两个通道主要是为了让这两个区域之间形成牢固的联系。

这座34 000平方米的建筑物采用了预应力混凝土、钢铁、木材和玻璃等建造。让-米歇尔·威尔莫特说：“我们的数码化孵化器项目是一个真正的催化剂，它将独特、创新、充满活力的空间中的两个主要创造性源泉集中在了一起——将一位19世纪工程师的才华与原生肆意的新生代想象力融合起来。”

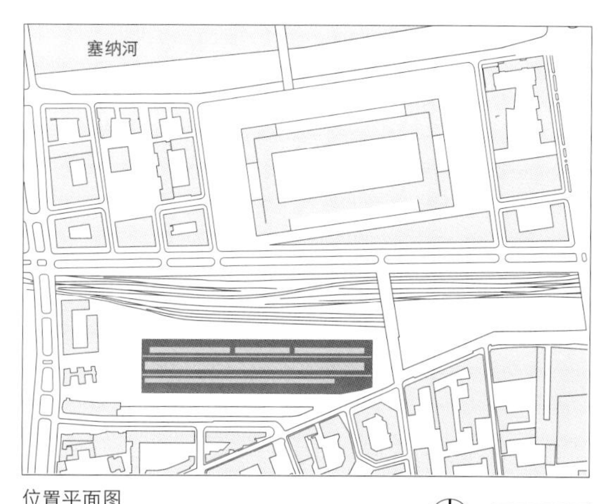

位置平面图

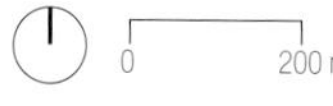

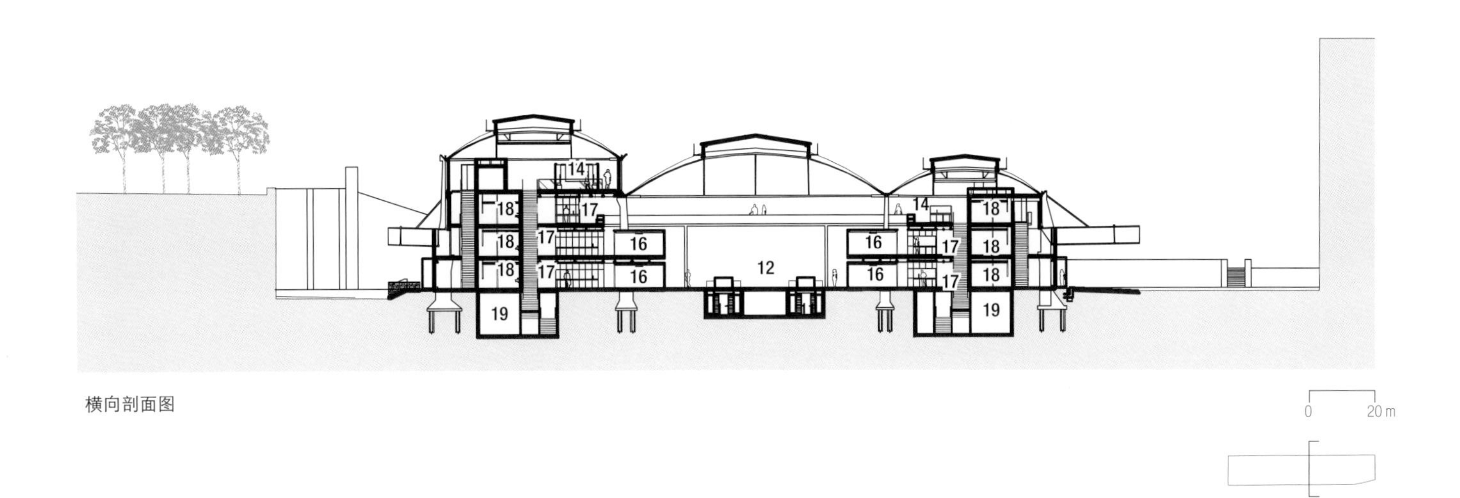

横向剖面图

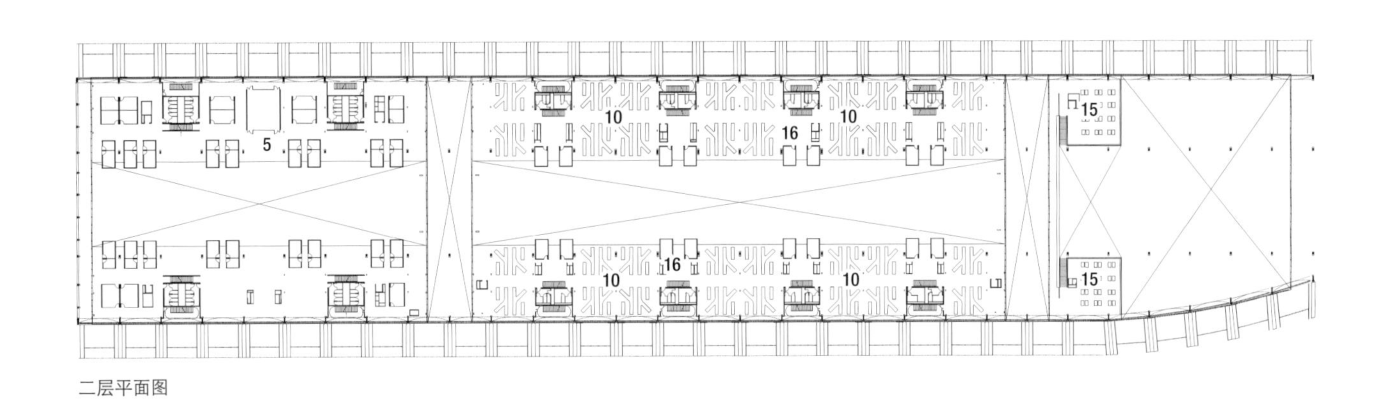

二层平面图

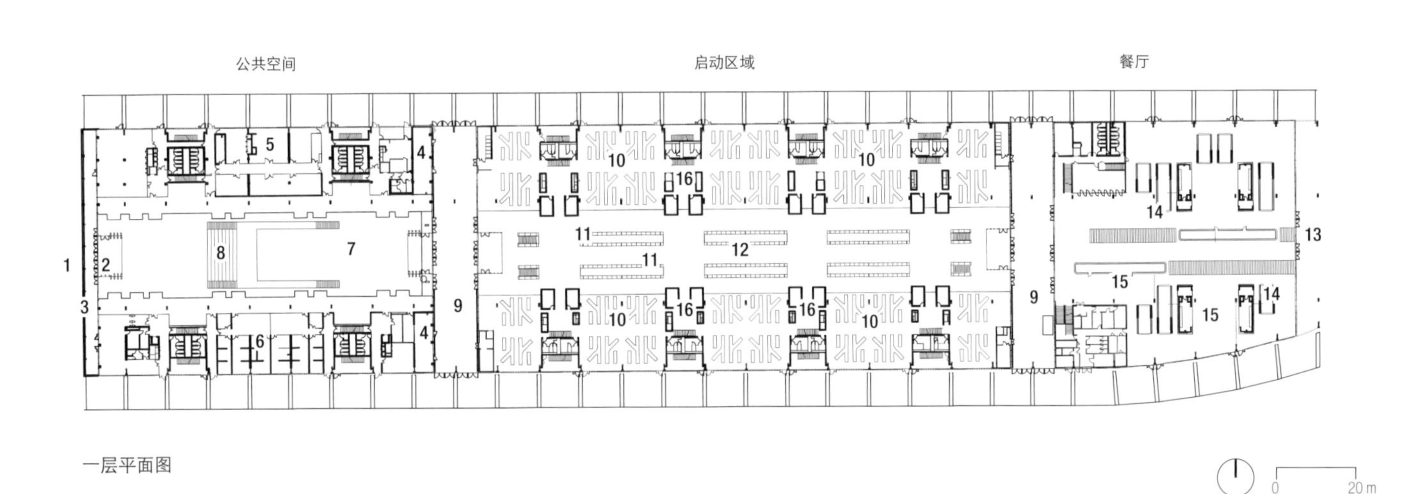

一层平面图

1 西入口
2 接待处
3 柱廊
4 商店
5 创客工厂
6 公共服务区
7 多功能空间
8 有360个座位的礼堂
9 公共通道
10 联合办公空间
11 衣帽间
12 收发点与多用途空间
13 东入口
14 厨房
15 餐厅
16 共用会议室
17 非正式会议室
18 洗手间
19 服务设施

PEGASUS

Mediaco
Mediaco
Mediaco

TATION

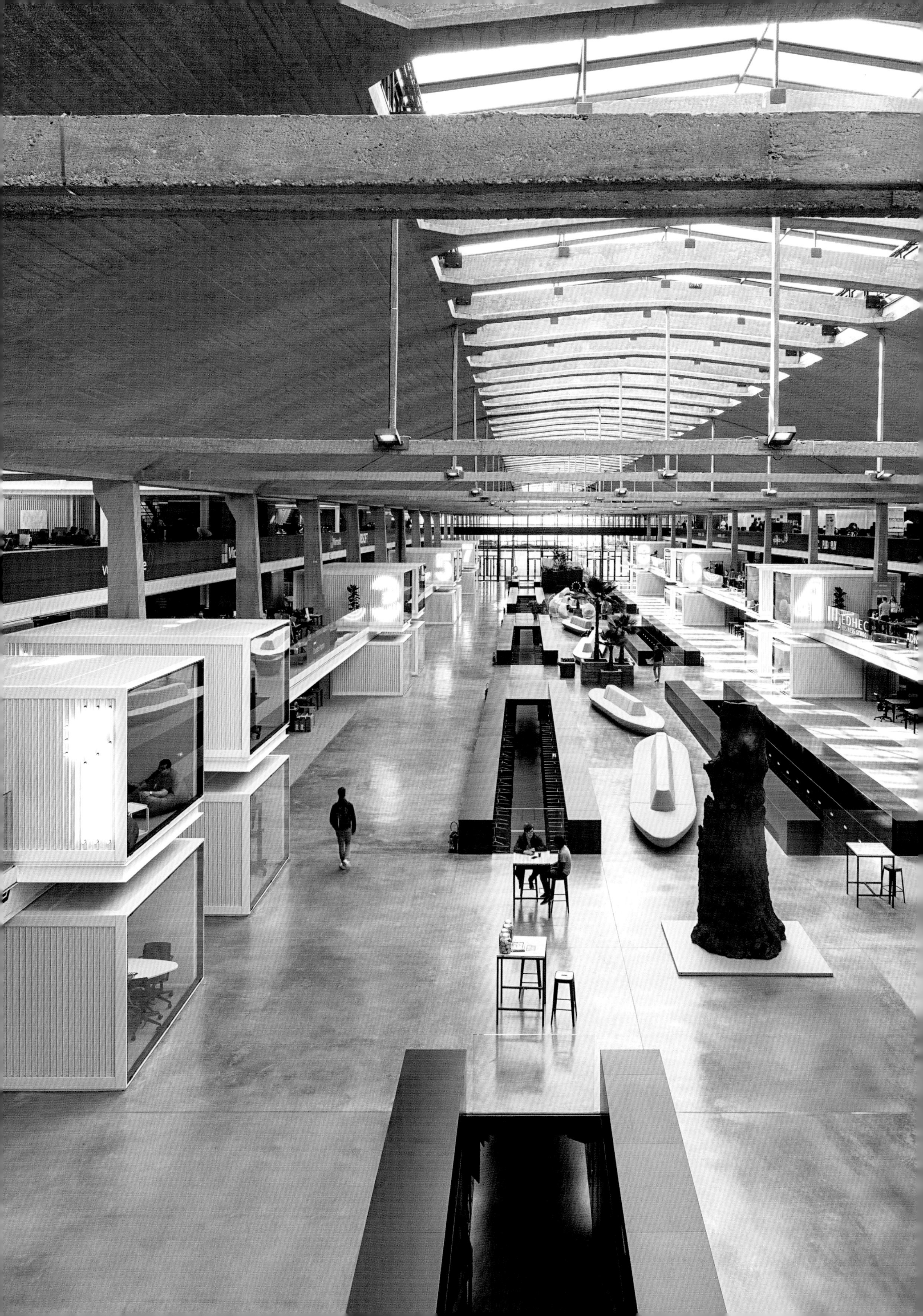
EDHEC

圣伯纳丁学院

这个雄心勃勃的项目涉及5000平方米的石制建筑翻新工程，这些建筑可以追溯到13世纪或15世纪。建筑师为建筑增加了一个特制的陶瓦新屋顶，用钢架支撑两层上层楼面，并在地下增加了300个混凝土微型桩以稳定这座旧建筑。

圣伯纳丁学院的历史可以追溯到1247年，是与巴黎圣礼拜教堂同一时期建造的。它是巴黎现存最大的非宗教性中世纪建筑，也是一个研究西方思想的重要中心之所在。法国大革命时期，这座建筑物不仅被剥夺了教育功能，也被改变了：它的学院餐厅——一个有三条过道的大厅被分隔开来，最终在1970年被遗弃，后来由消防队员驻扎在那里。

威尔莫特与历史建筑领域的资深建筑师埃尔韦·巴普蒂斯特一道，在巴黎教区协会的支持下对学院进行了翻新，复活了这座精致的哥特式建筑，通过光影的演绎使之更加富有生气。原来的礼堂是一个长达68米的完整空间，现在用作接待处与展览区以及一家书店。

这个项目必须在陡峭的倾斜屋顶结构下创建出足够大的使用面积，而且传送到地面的负载必须是受到严格限定的。建筑师的办法是利用一对大门桁架，同时也将其兼做各种服务设施的围护结构，从而达到进一步加固内部空间的效果。改造后的学院现在拥有两个礼堂（原本并不存在）。设计方案将较低楼层的天花板悬挂在建筑结构的金属连接梁上，这样减少了礼堂中细柱上的负载。该项目赢得了2010年欧盟文化遗产奖。

项目地点 / 法国巴黎
完工时间 / 2008年
建筑面积 / 5000平方米

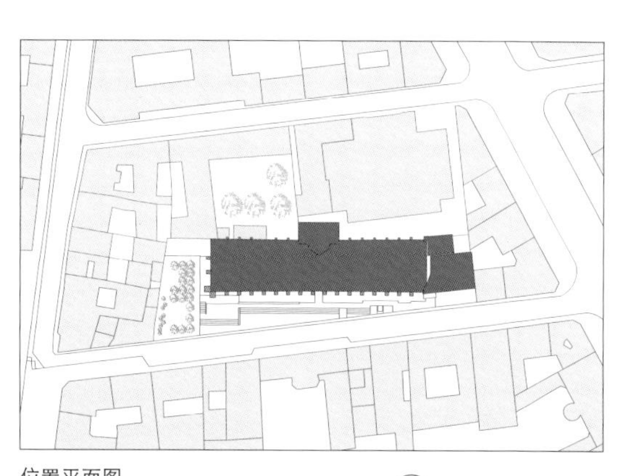

位置平面图

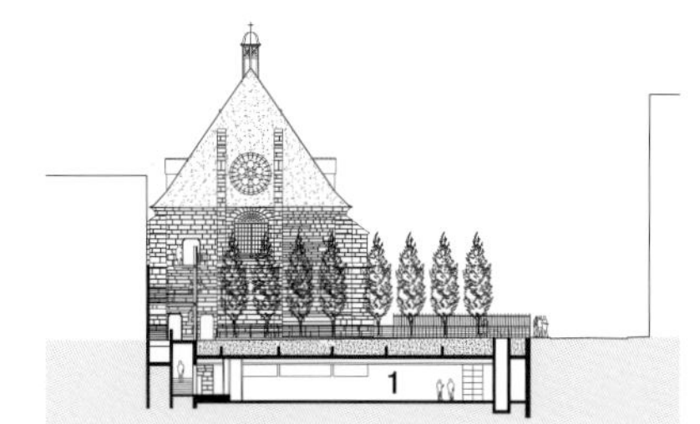

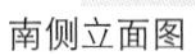
南侧立面图

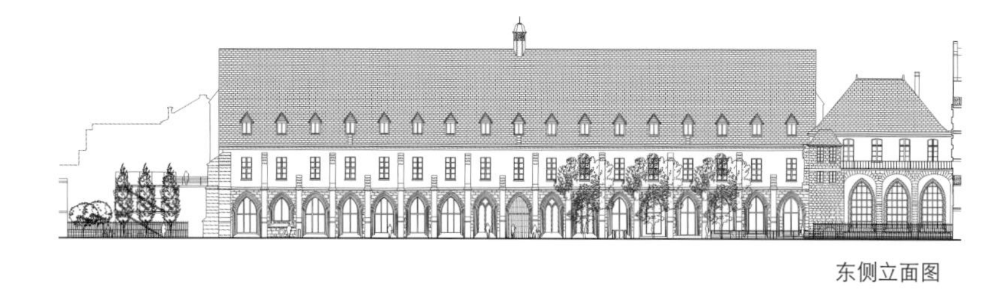
东侧立面图

横向剖面图

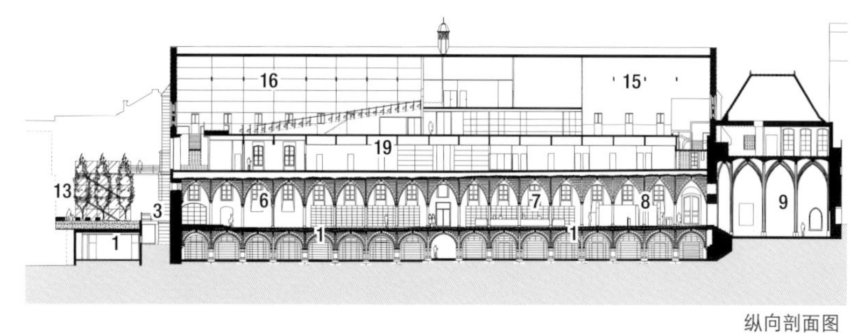

纵向剖面图

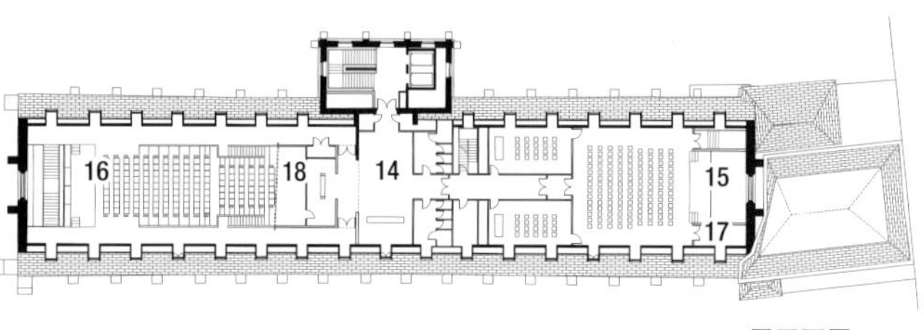

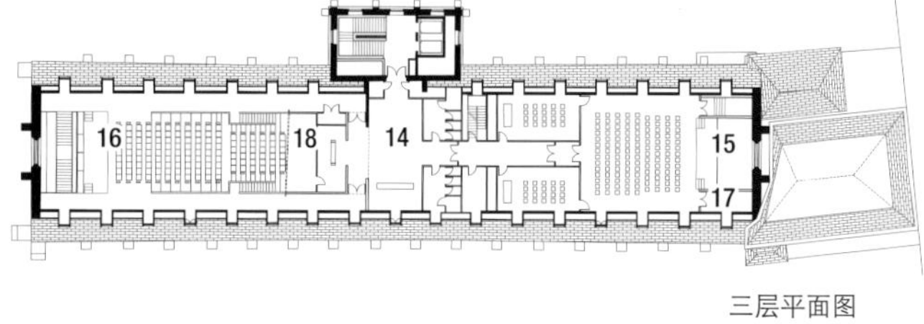

三层平面图

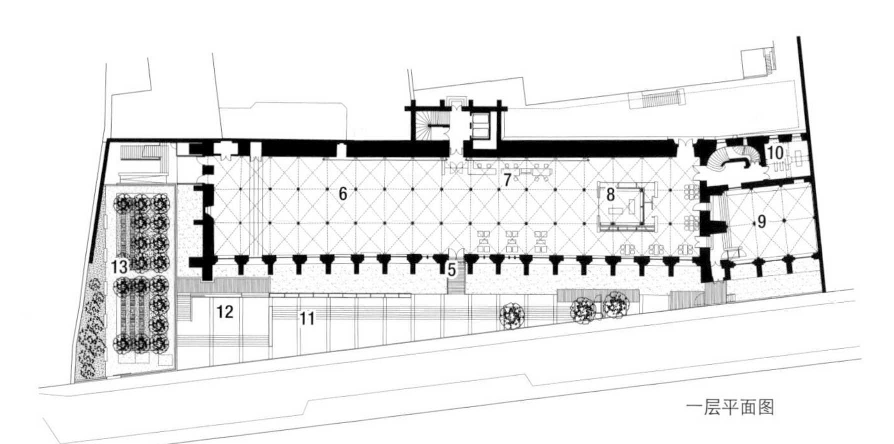

一层平面图

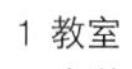

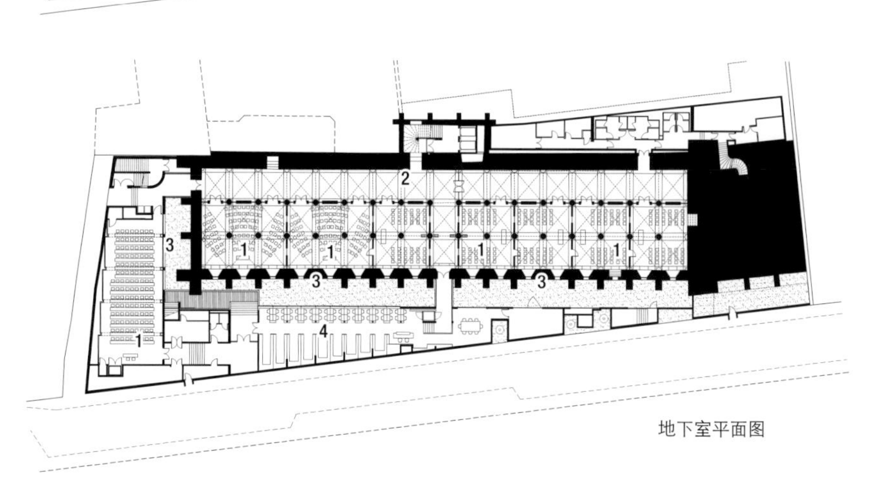

地下室平面图

1 教室
2 步道
3 排水渠
4 图书馆
5 入口
6 展厅
7 接待处
8 书店
9 音乐室
10 祷告室
11 上层前庭
12 下层前庭
13 花园
14 大厅
15 小礼堂
16 大礼堂
17 档案室
18 座位下的服务设施空间
19 办公室

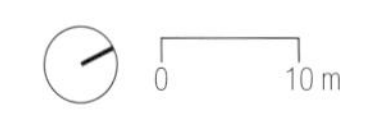

URBAN PROJECTS
城市规划项目

大莫斯科
城市规划方案

项目地点 / 俄罗斯莫斯科
方案时间 / 2012年
建设面积 / 668平方千米

2012年初，莫斯科市启动了大莫斯科城市规划方案国际招标，规划面积为668平方千米，是现在城市规模的1.5倍，这个方案将用于这座拥有1200万人口的扩城建设。

俄罗斯首都发展战略方案由十个团队参与提供，其中“法国巴黎方案”普遍被看好。这些研究来自于2012年1月至8月每个月举办一次的研讨会。2012年9月5日，一个评审委员会授予由城市建筑师安东尼·格伦巴赫、让-米歇尔·威尔莫特和谢尔盖·特卡琴科领导的法俄团队进行莫斯科扩城规划的设计任务。

法俄团队规划方案的指导原则之一是围绕莫斯科河发展大都市区的公共机构与休闲活动设施建设，就像巴黎的塞纳河和伦敦的泰晤士河一样，莫斯科河也将因此成为这座城市特征的载体。随着对生活质量与原有城市遗迹价值的持续关注，以及这里如今对于行人而言存在着出行难的困惑，因此莫斯科河河岸的景观开发将与公共交通的重新部署（在这里体现为有轨电车）相伴而生，以限制日益增长的交通流量。

今天，莫斯科市经常因缺乏公共交通设施而造成交通瘫痪。到2025年，莫斯科将有1400万名居民，这也是中标团队提出一个具有全球视野的大都会公共交通方案的原因。根据该规划，城市需要建设多条通往机场与莫斯科郊外的快速列车线路（MKAD），而交通堵塞问题则应该通过地铁网和有轨电车线路等多种运输方式来缓解。新的特快地铁线路——新莫斯科线（NML）将有效地将旧城与大莫斯科地区联系起来。虽然大莫斯科项目的主要目的是扩城，但格伦巴赫和威尔莫特认为重构现有的城市发展脉络比建设新城区要重要得多。

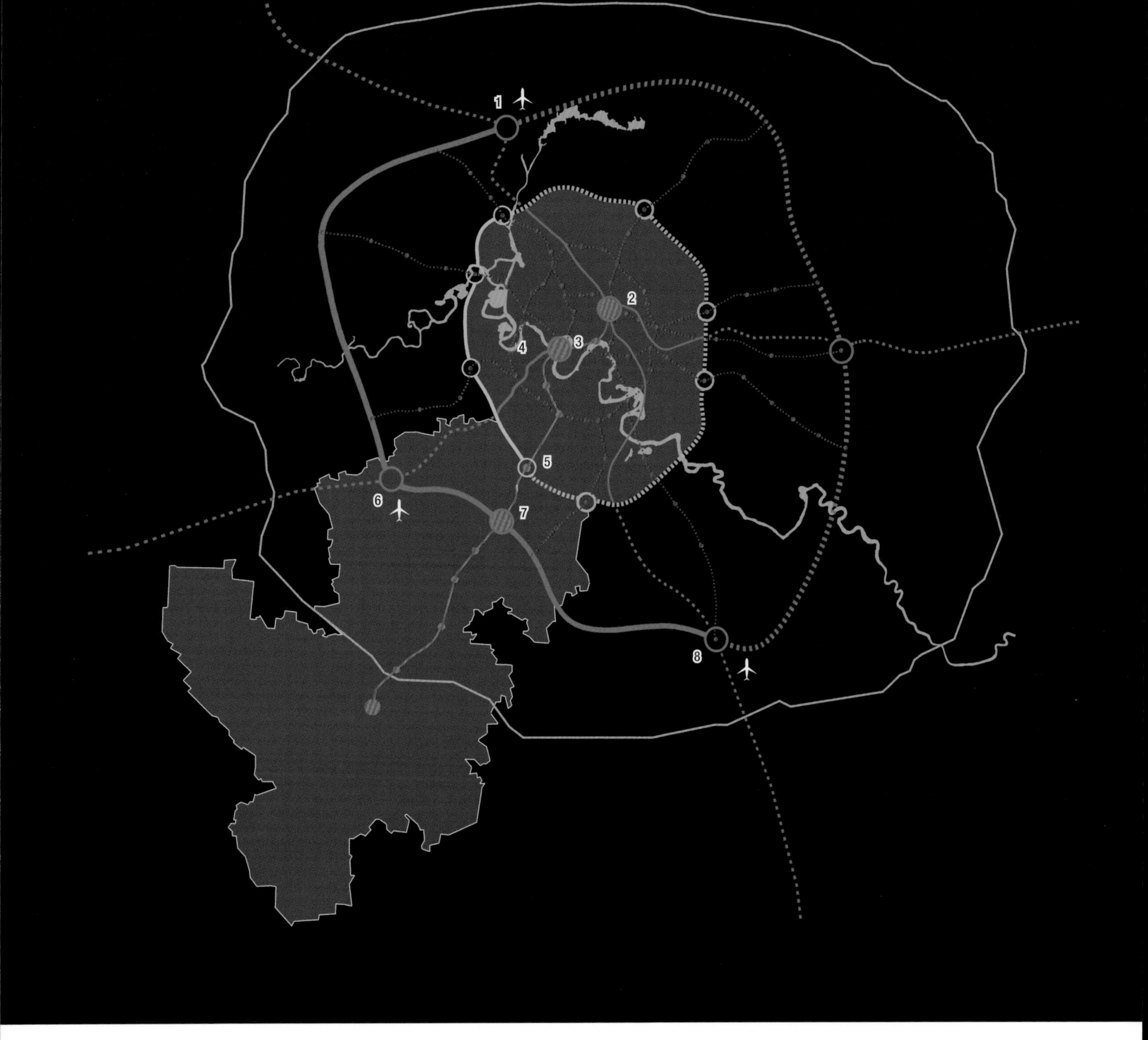

反思城市交通

0 20 km

- 莫斯科公共交通网各大车站
- 高速列车/ MKAD莫斯科环城公路及周边地区公路
  - 第1期：2020—2025
  - 第2期
- 高速列车
  - 第1期：2020—2025
  - 第2期
- MNL地铁快线：2020—2025 / NML高速地铁：2020—2025
- 地铁
- 莫斯科环城公路
- 新莫斯科延长线

1 莫斯科谢列梅捷沃国际机场站
2 勒斯特洛斯站
3 克里姆林宫站
4 基辅站
5 新莫斯科站
6 伏努科沃国际机场站
7 科穆纳尔卡站
8 莫斯科多莫杰多沃机场

新莫斯科之门（改造前和改造后）

沿着克林姆林宫方向的莫斯科河河岸（改造前和改造后）

三座车站将整座莫斯科城连结为一体（改造前和改造后）

基辅斯基火车站（改造前和改造后）

让–饶勒斯大街项目

项目地点 / 法国尼姆
完工时间 / 2013年
建设长度 / 1.6千米

尼姆最大的街道，现在名为让–饶勒斯大街，是1753年由军事工程师雅克·菲利普为路易十五国王设计的，用于将来自尼姆喷泉花园中的水引导下行到一条1.6千米长、61米宽，种植着四排欧洲荨麻树的漫步大道附近。不幸的是，它的吸引力逐渐随着一个存在乱停乱放现象的停车场而受到不断地侵蚀。

由于当地居民想要重获让–饶勒斯大街原来的非凡风貌，因此威尔莫特提出把它变成一个愉快的聚会场所。它的功能如同尼姆喷泉花园的一个延伸部分，这条中央步行街在枝繁叶茂的绿树之下石路绵延，有着多种多样的用途：有操场，有集会场所，还有临时摊位等。让–米歇尔·威尔莫特设计了亭子与绿廊装点着宽阔的让–饶勒斯中央步行街，利用古老茂密的荨麻构成了各种图案。两侧的外墙由采用LED背光照明的冲孔薄铁板组成，通过提供普罗旺斯花边样式的精致照明使这条大街变得生气勃勃。“我们想将尼姆喷泉花园的活力引到大街上，让人们在那里也可以感觉惬意，会更乐意去那里漫步。”建筑师说。

这条大街始于有一系列观赏性水池的花园脚下，它已经成为迅速被尼姆居民们普遍接受的、令人愉快的景点。项目所使用的材料包括表面喷砂处理混凝土（人行道）、混凝土板、石板、克罗地亚天然石材制作的铺路石（中央步行街）和木制铺板（阳台）等。

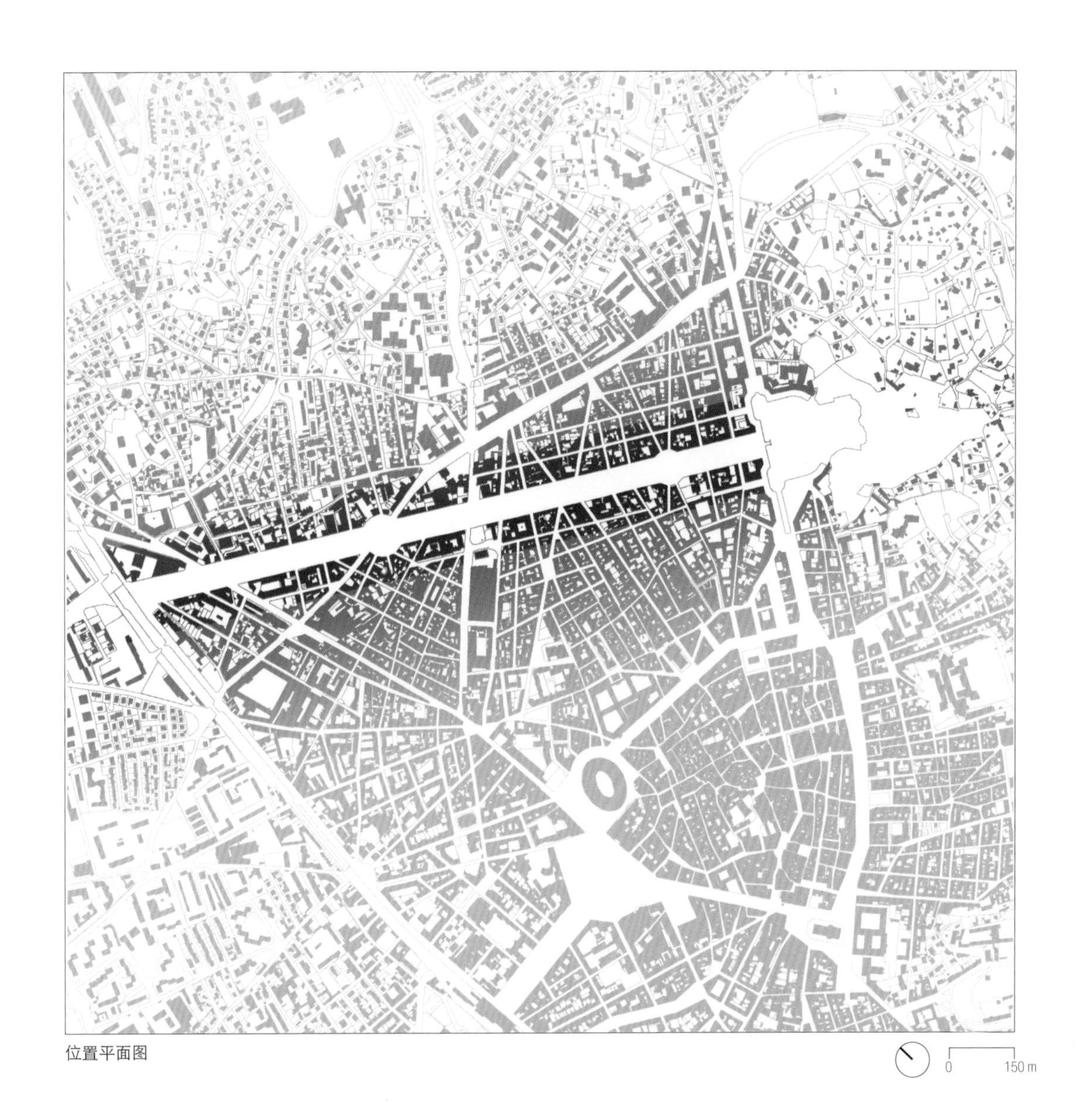
0
150 m
位置平面图

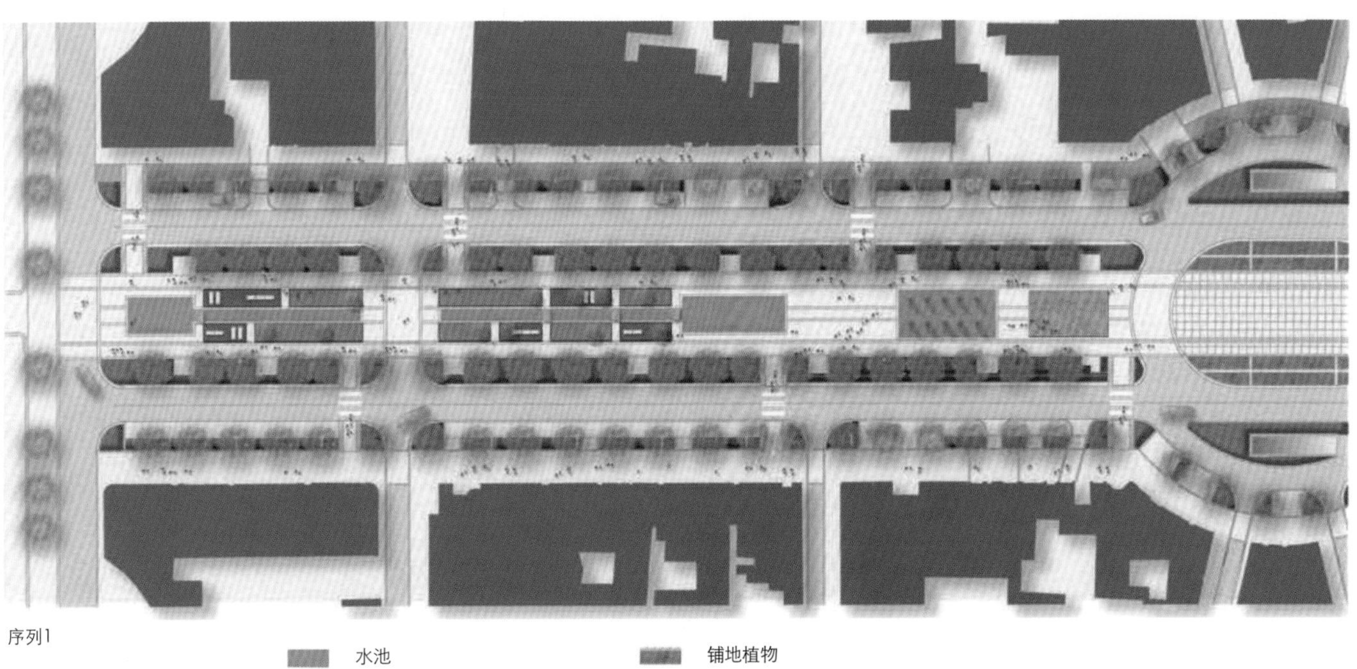
水池
旱式喷泉
铺板和长凳
草坪
铺地植物
路面：混凝土与铺路石
成熟的朴树
灌木
0
20 m
序列1

MUSEUMS
博物馆项目

卢浮宫原始艺术厅

项目地点 / 法国巴黎
完工时间 / 2000年
建筑面积 / 1400平方米

在总统雅克·希拉克和部落艺术品经销商雅克·科切克的推动下，法国政府和卢浮宫共同努力为“原始艺术”在这座博物馆中找到了合适的位置。

让–米歇尔·威尔莫特是原始艺术厅重新设计招标项目的胜出者，这个艺术厅位于塞纳河畔的卢浮宫二楼。“卢浮宫原始艺术部的这件作品”，威尔莫特说，“代表了以我们的方式去进行博物馆设计的一个分水岭——这意味着我们要最大限度地简化展览空间以便让艺术作品能够脱颖而出。”

2000年4月13日，原始艺术厅作为卢浮宫的一个分支机构（原始艺术部）以及对未来的布朗利河岸博物馆的“预展”正式亮相，然后，还要将其建设成为重组几个法国原始艺术博物馆的综合性展厅。

这个约1400平方米的展览空间致力于展出雅克·科切克精心挑选出来的非洲、亚洲、大洋洲和美洲的原始艺术品。这是“原始艺术”第一次在以西方艺术为主导的卢浮宫中展出。威尔莫特的团队关心的是，即使原始艺术厅的设计适应了展厅的各种作品，尊重了这座建筑的悠久历史，但仍然要与卢浮宫其他部分保持一种适当的联系。

轻柔灯光下的原始艺术厅为参观者们提供了干净、开放、分隔、简化的展柜和有所限定的空间分区。极简主义的展柜根据展厅的具体情况而构思设计，并且根据展品尺寸而刻意留出宽松的空间，这样有助于为展品营造出一种神圣感，参观者也可以欣赏到每件雕塑的完整维度。这也是威尔莫特第一次采用这种基于最佳透明度的方法，展厅中的照明设施不再像以往那样纳入展柜中，而是采用了一种灵活的高架式系统，它能够在房间的任何地方创建光源。

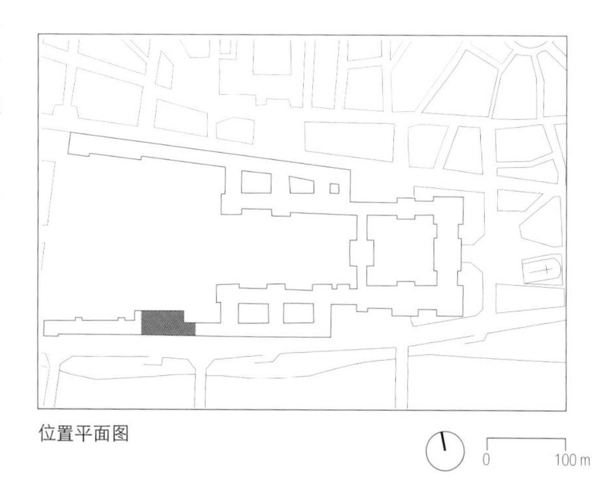

位置平面图

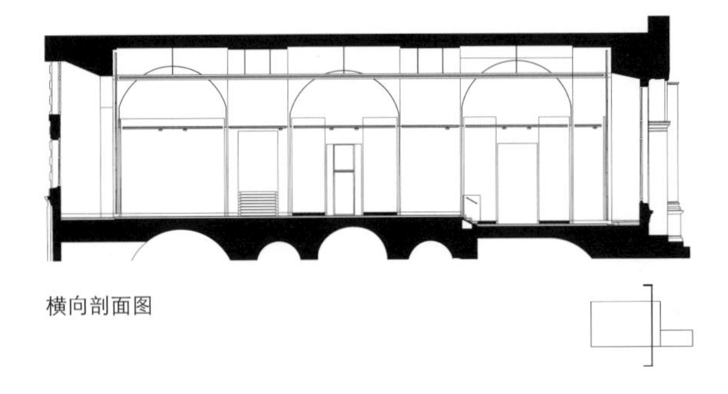

横向剖面图

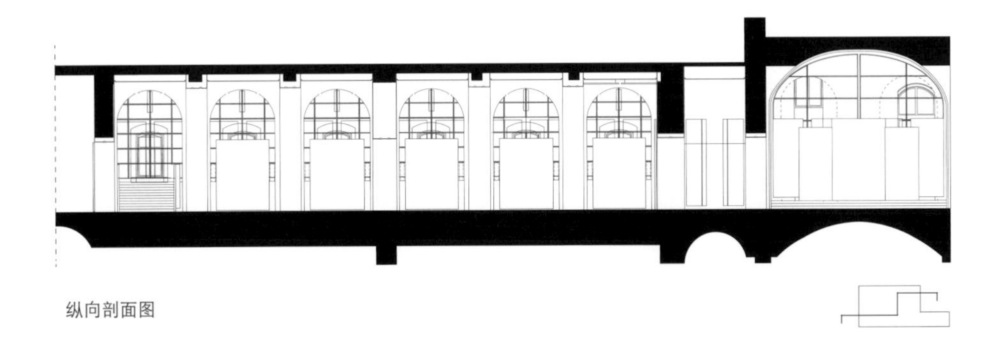

纵向剖面图

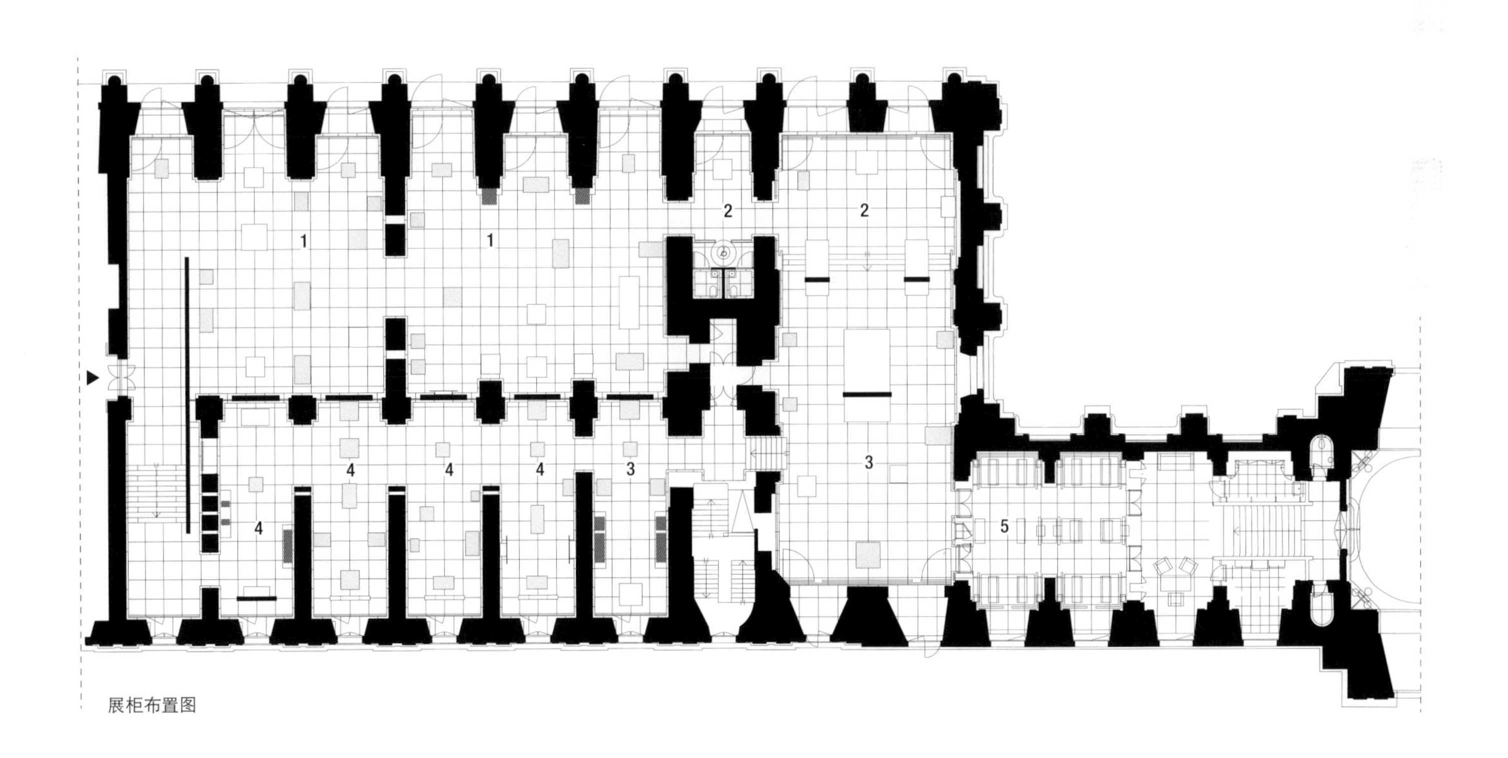

展柜布置图

1 非洲原始艺术品区
2 亚洲原始艺术品区
3 大洋洲原始艺术品区
4 美洲原始艺术品区
5 多媒体区

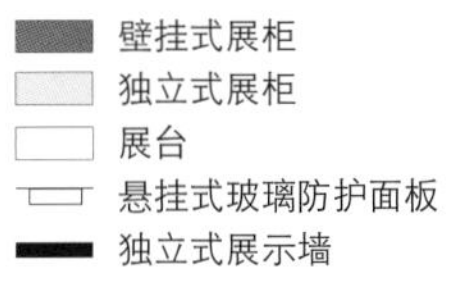

Océanie

佳娜艺术中心

让-米歇尔・威尔莫特在韩国的首个作品——佳娜艺术中心，面积为3024平方米。这座钢筋混凝土建筑致力于当代艺术展览，外覆层使用了意大利米黄石灰石和木材。项目客户是艺术品经销商李浩宰。该项目还包括一家精品店、一家餐厅和一个露天剧场。

在首尔北部平昌洞山丘上的一个偏远住宅区，佳娜艺术中心（佳娜艺廊）建在露天剧场上方的斜坡上。这是个四层楼的结构，有一部分在地下。人工照明在艺廊中大量使用。在下沉的圆形剧场上方，阳台的玻璃走廊使如画般的风景渗透到了建筑的中心，而艺廊则从树木繁茂的山坡上铺陈而下，其外表则是一圈观景木台。

佳娜艺廊的第一座建筑一年就建成了，接着在邻近地点建造了一座铜制覆层的拍卖室，与艺廊建筑的浅色石材相映成趣。三年后，距离艺廊一条街道的地方增建了第三座建筑，使之成为一个艺术品存放点，这个三层建筑还包括一个有一定角度的展览空间。

青砖是建筑师最喜欢的材料之一，它与深色大窗格玻璃形成了一种东方式灯笼的效果。几座建筑物都在彼此的视线之内，似乎可以通过它们各自的设计元素将彼此连接起来。在佳娜艺廊完工后不久，这位客户又提出了一个早已有之的想法，就是在首尔市中心附近的文化艺术街仁寺洞建造一座艺术中心。

建筑地点 / 韩国首尔
完工时间 / 1998年
建筑面积 / 3024平方米

位置平面图

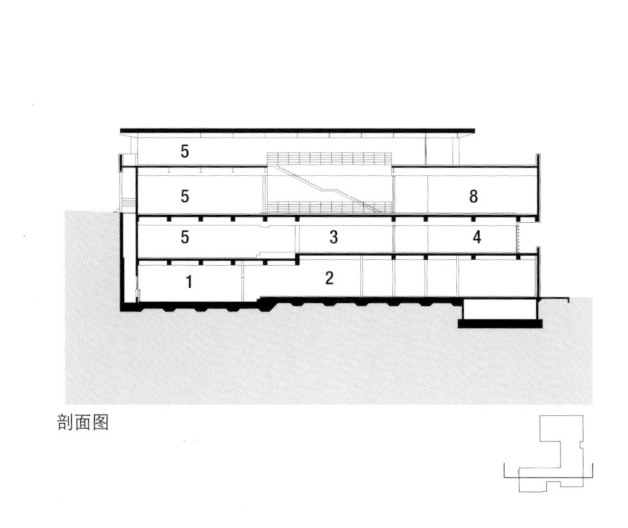

剖面图

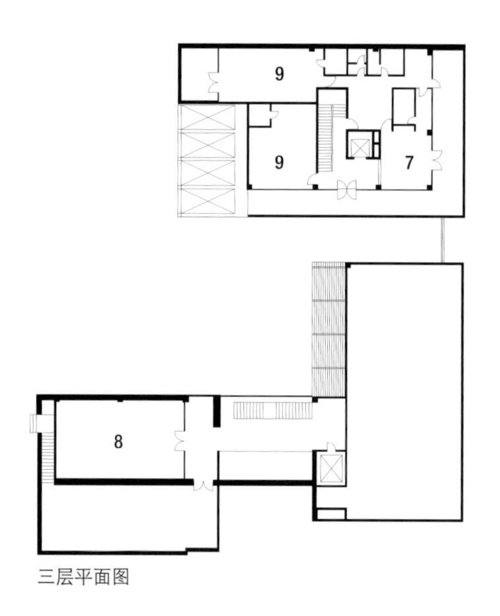

三层平面图

一层平面图

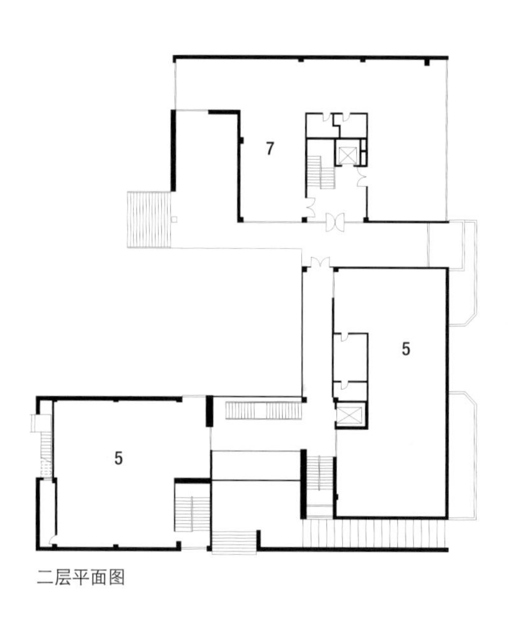

二层平面图

地下室平面图

1 艺术品储藏区
2 车库
3 主入口
4 咖啡厅
5 常设展览区
6 礼品店
7 办公室
8 临时展览区
9 卧室

雅克·希拉克总统博物馆

项目地点 / 法国萨尔朗
完工时间 / 2002年（第一期），2006年（第二期）
建筑面积 / 5267平方米

这个项目涉及翻新、室内设计和博物馆的后续扩建工作，完工后打算用来展出雅克·希拉克在他两届法国总统任期内收到的6500份礼物中的约200件珍品（1995—2007）。这座博物馆位于总统的家乡——法国西南部的科雷兹地区。这座钢筋混凝土建筑的外表以科雷兹特有的粉红色花岗岩、橡木和天然石板屋顶为特征。项目第一阶段的建筑面积为1500平方米，2002年第二阶段再增加3767平方米。

让-米歇尔·威尔莫特回忆，对这里经过两个小时的参观访问后，他就找到了行动方案，便没有再犹豫。这个博物馆背对着一幢小径木建造的大型农舍式建筑群，应该成为与此处景观相适应的田园式作品。一块沼泽地由景观设计师米歇尔·德维涅改造成了草地，这里原来有一个大池塘被中空的芦苇丛包围着。威尔莫特解释说："最初的轮廓只不过是提出了在这里建一座博物馆的想法而已。"因此，他的任务是在这里被指派一名馆长之前从形式和内容上发展这一理念。

展览要陈列的展品需要根据希拉克总统的地缘政治情况进行选择，要区分五大洲并遵循地理情况和时间顺序进行安排布置。在第一阶段，两个主要建筑物的形式与规模受到了以前伫立在入口处的建筑物的启发，博物馆本身占据了第二座建筑，而第一个建筑专门用于临时展览。第二阶段则要继续建造类似的第三个建筑体，与之前的一致，是一座在这个场地上的中世纪塔堡，作为这座博物馆的一个标志矗立在增建的阳台尽头，在新的仓库空间之上吸引着人们的目光。这个塔堡被用作向公众开放的图书馆，围绕着中空的内庭建造，有三层的高度。

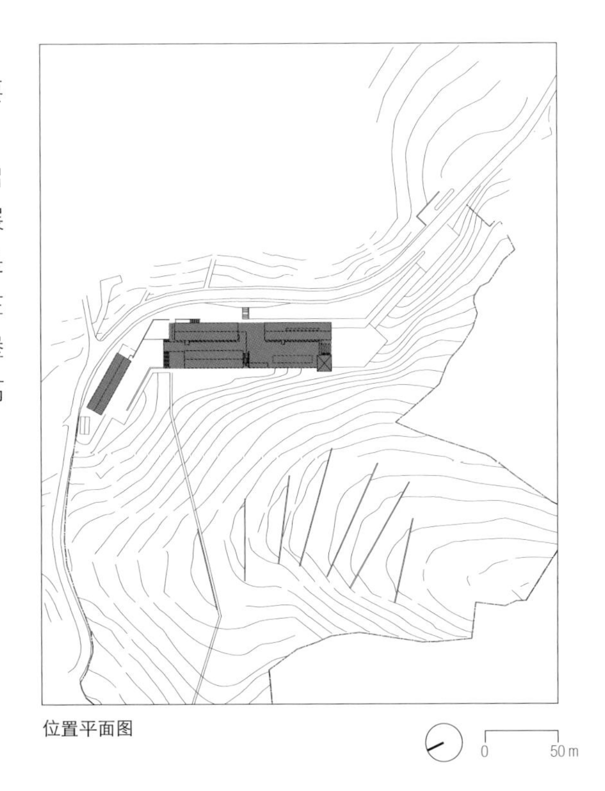

位置平面图

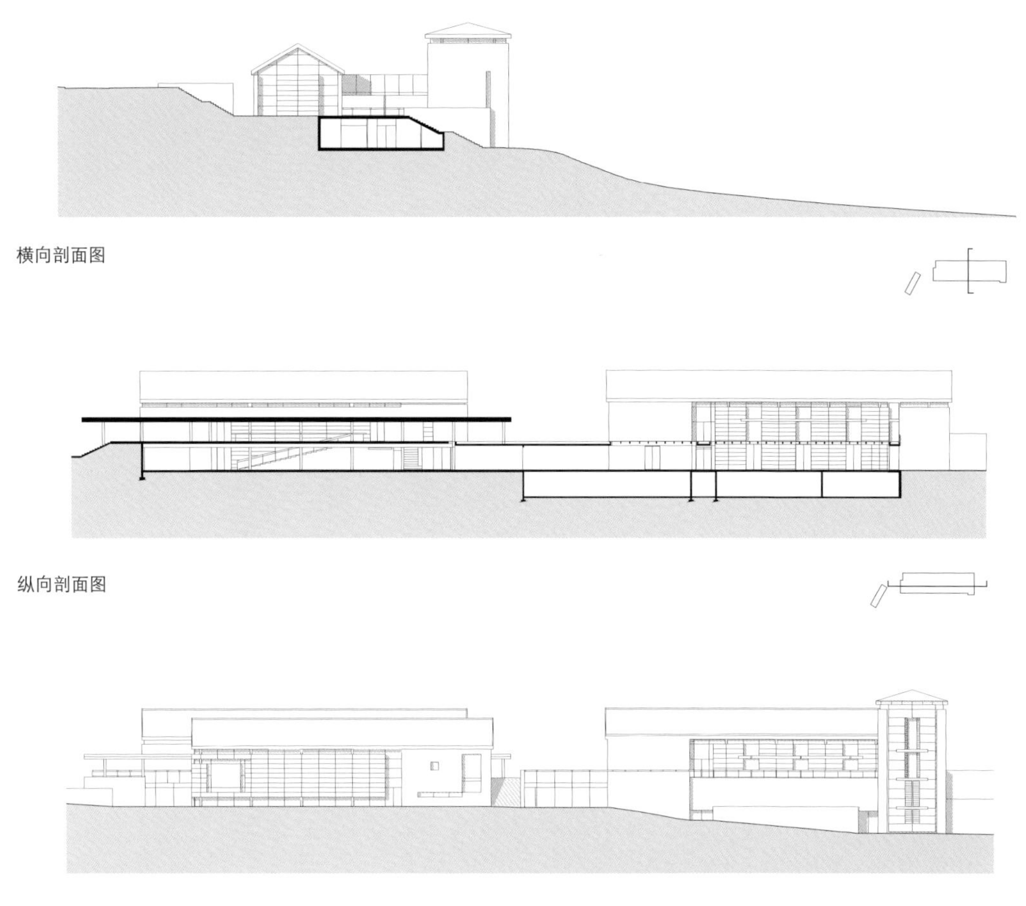

横向剖面图

纵向剖面图

西侧立面图

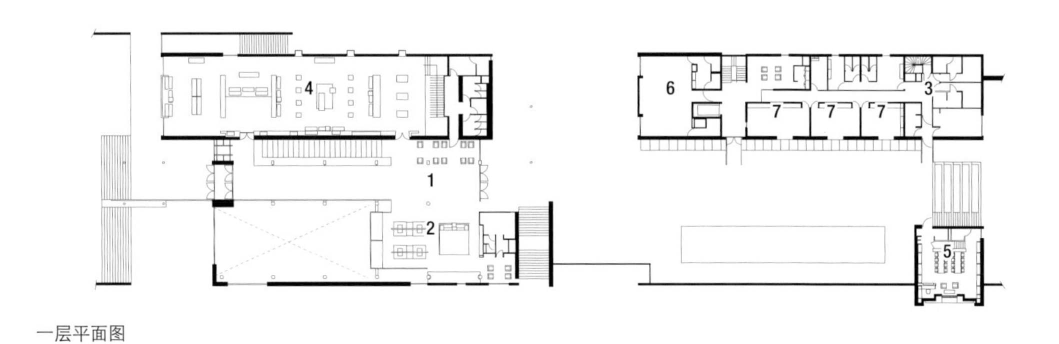

一层平面图

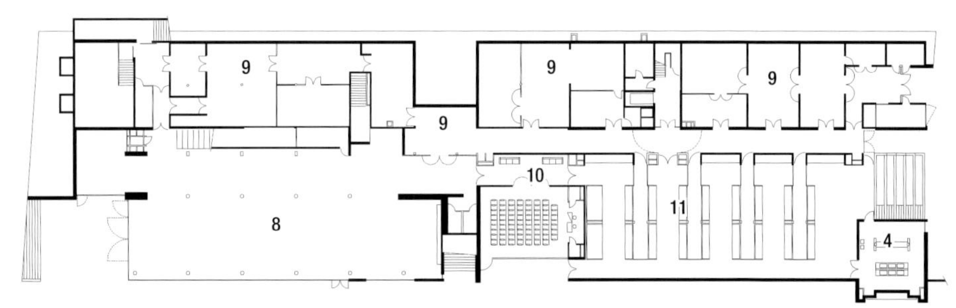

花园层平面图

1 入口
2 书店
3 衣帽间
4 常设展览区
5 图书馆
6 活动室
7 办公室
8 临时展览区
9 展品维护工作室
10 会议室
11 公共寄存处

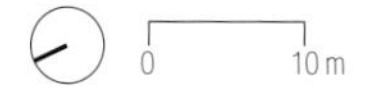

photographies
de personnalités
politiques

尤伦斯
当代艺术中心

项目地点 / 中国北京
完工日期 / 2007年
建筑面积 / 7500平方米

这是个艺术画廊项目，让–米歇尔·威尔莫特修复并翻新了这座20世纪50年代，面积为7500平方米的工业建筑。建筑本身是钢筋混凝土的，建筑师还为其增加了质地优良的环氧树脂地板和彩绘墙壁。这个艺术中心包括一个大型展厅，面积达2000平方米，其余的空间则作为书店、酒吧、食堂、图书文档中心、VIP空间、餐厅、有135个座位的露天剧场和280平方米的办公空间等。这里的天花板高达9.6米，足以为现代艺术展示提供所需要的所有空间。

盖伊·尤伦斯和米莉恩·尤伦斯夫妇是艺术爱好者和收藏家，他们收藏了相当数量的中国艺术品，规模是世界上同类艺术品中较大的。超过1300件的艺术品由几代艺术家创作，其中包括雕塑、绘画、装置和视频等。这个中心位于北京大山子798艺术区，曾经是一个工业厂区。

尤伦斯夫妇基金会于2002年在瑞士成立，他们是中国艺术界的积极支持者。基金会赞助了世界各地的中国艺术活动和展览，其中2003年和2005年的威尼斯中国艺术双年展基金会从其收藏品中借出了一部分给博物馆和艺术中心并组织了展览。

威尔莫特基于这座建筑的内在特性进行了改造，去除了原建筑的一些东西之后，又增加了一些新的元素。各处7.4米高的门框和一个34米高的砖砌烟囱构造出了这个高大的空间。这个综合体的结构是基于两个大型殿堂式的、并排设置的空间。在第一个区域，艺术中心入口、书店和食堂、露天剧场以及200平方米的较小展馆占地面积总共为280平方米。图书文档区和VIP区则位于夹楼。第二个空间有非常大的开放式表演和展览区域，布局上尽可能不受任何约束。

头顶的自然光沿着中轴线扩散与人造光联系在了一起，在这个空间中产生了一种均匀的亮度。中心的整个配置情况符合艺术展览的国际最高标准。

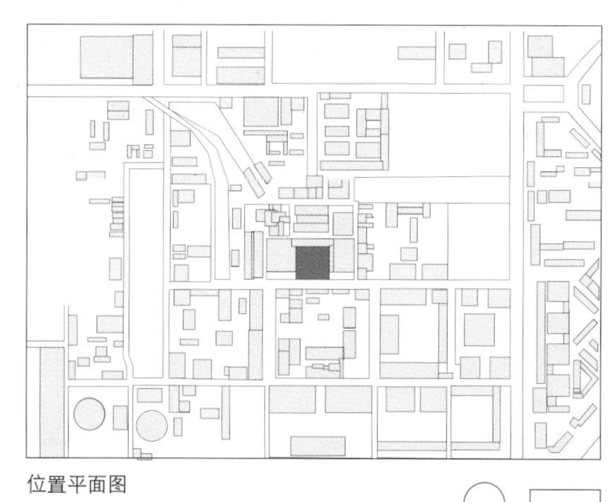

位置平面图

0 200 m

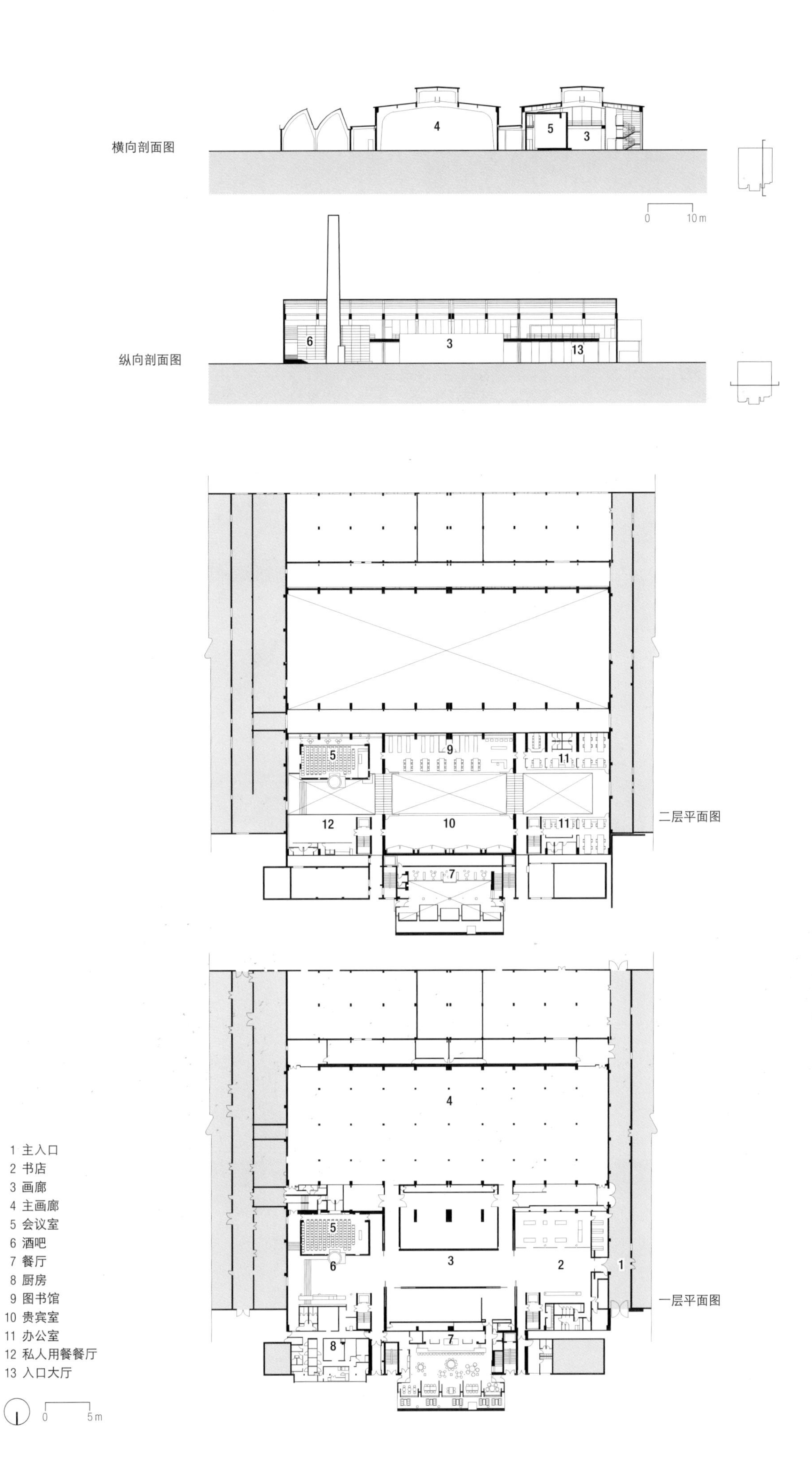

1 主入口
2 书店
3 画廊
4 主画廊
5 会议室
6 酒吧
7 餐厅
8 厨房
9 图书馆
10 贵宾室
11 办公室
12 私人用餐餐厅
13 入口大厅

多哈伊斯兰艺术博物馆

这是让–米歇尔·威尔莫特与贝聿铭合作的博物馆项目。伊斯兰艺术博物馆位于卡塔尔西海湾对面滨海大道尽头的人工岛上，威尔莫特负责室内展览画廊的设计工作。这两位大师以前曾在巴黎卢浮宫项目上有过合作。

因为客户经常扩展博物馆的各类收藏品，所以这里的展览设计必须尽可能灵活。占地面积达5250平方米的画廊以一种规则的几何图案组织起来，从一个画廊到另一个画廊的视野很开阔。这里保持着贝聿铭设计特点中组织安排的纯净风格。每个画廊在墙上有许多展柜，这样可以留出中央空间以将艺术杰作安放在定制的展柜中。

威尔莫特建筑事务所设计的展柜因其根据展品尺寸大小定制，表现出了高度的技术性能，所以是独一无二的。70种不同类型的展柜非常大，可以容纳任何类型的物件或一组较小的展品。玻璃展柜没有垂直的金属框架，所以它们在展品周围似乎是不存在的。在展柜内，每个展品都是在中央舞台上聚光灯下的“明星”。展品看起来似乎是在空中漂浮着的，这是通过周围的暗色与精心设计的光纤照明方案而营造出来的效果。而建筑材料采用的珍惜木材与石材，更加增强了这种效果。深色的巴西悬铃木和斑岩石为了突出展品而隐藏了起来。同时，它们纹理的丰富细节比之其平整的表面来说反而能使展柜更为突出。而且，这些材质的排列方式在每个画廊中也各不相同，可以使参观者很容易地找到方向。

该方案还包括配置了非常灵活的700平方米的临时展馆。书店是作为贝聿铭设计的中庭的延续部分。展柜和墙壁内隐藏着一些小壁龛。威尔莫特专门为伊斯兰艺术博物馆设计了家具，提供了公共场所的桌椅和办公家具等。家具选用了木材、皮革和金属等经久耐用的材料。

项目地点 / 卡塔尔多哈
完工时间 / 2008年
建筑面积 / 5250平方米

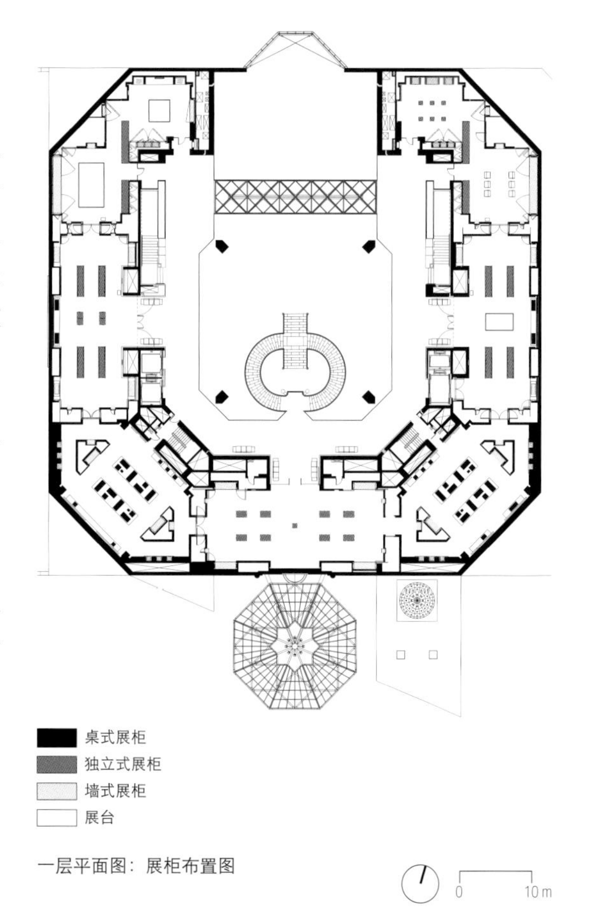

一层平面图：展柜布置图

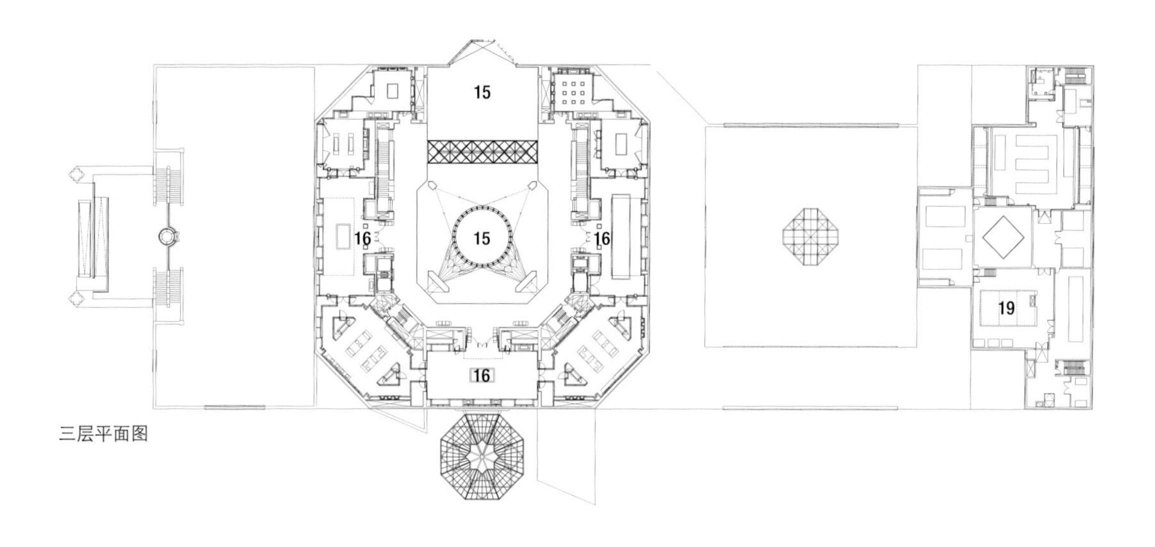

三层平面图

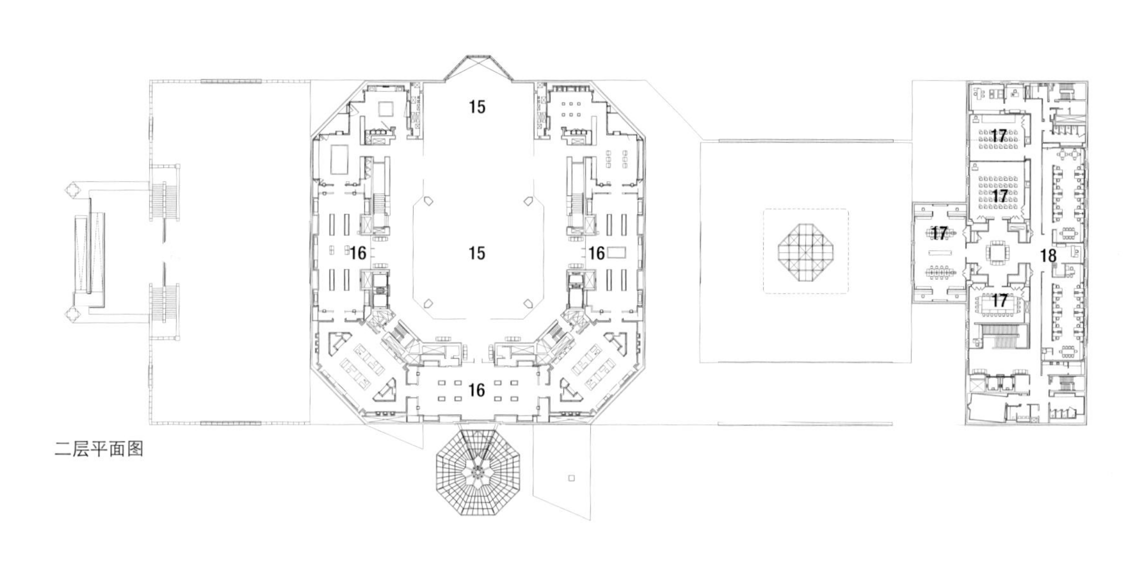

二层平面图

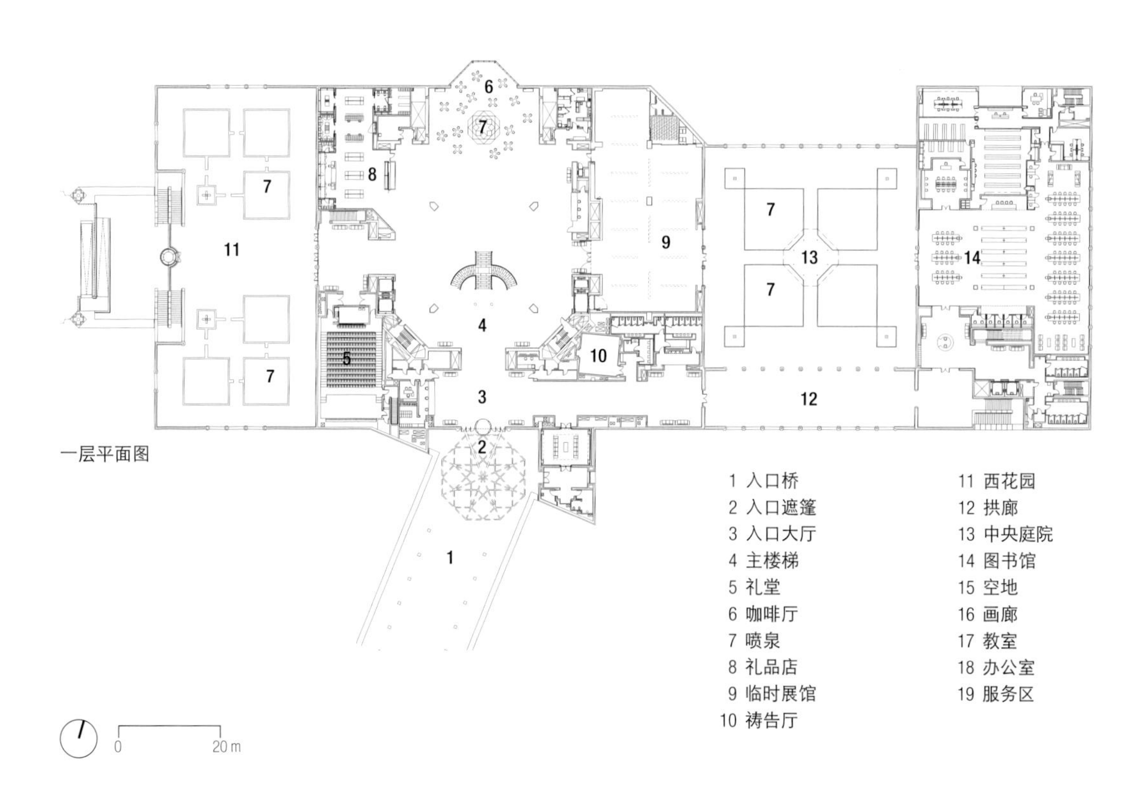

一层平面图

1 入口桥
2 入口遮篷
3 入口大厅
4 主楼梯
5 礼堂
6 咖啡厅
7 喷泉
8 礼品店
9 临时展馆
10 祷告厅
11 西花园
12 拱廊
13 中央庭院
14 图书馆
15 空地
16 画廊
17 教室
18 办公室
19 服务区

奥赛博物馆
翻新改造项目

项目地点 / 法国巴黎
完工时间 / 2012年
建筑面积 / 2200平方米

让–米歇尔·威尔莫特在2009年的奥赛博物馆各区域及书店礼品店翻新招标项目中中标，项目涉及的总建筑面积为2200平方米，包括这座巴黎著名的博物馆中5楼1233平方米的印象派画廊，以及位于同一层的临时展馆。奥赛博物馆，坐落在建筑师维克托·拉鲁于1898年到1900年间设计的同名火车站内，拥有世界上最大规模的印象派艺术收藏品。

项目的设计理念是需要对原建筑进行修复翻新工作，处理每件作品上的自然光线，打造出更为顺畅的参观通道，同时还要能够更为灵活、完整地收藏艺术品。尽管在自然光和人造光之间已经建起了对话，但威尔莫特还是试图保留原建筑那独特的风格特征。建筑师将两侧的天花板降低使之与人体高度相匹配，这样可以使人们与那些不朽的艺术作品亲密接触，它们在画廊中间上方原建筑结构引入的光线下展现在观众面前。

地上铺设了深色木制地板，而墙壁被描绘成精美的岩灰色，这可以使绘画作品色彩运用的细微之处更好地展现出来。而各种雕塑则在许多展柜中安静地伫立着，既被保护得很好，又显得十分突出。中央区域增加的间隔墙则用来展示关键的艺术品。

独立式展柜
服务设施管道与独立展位
展台
吉冈德仁设计的玻璃长椅

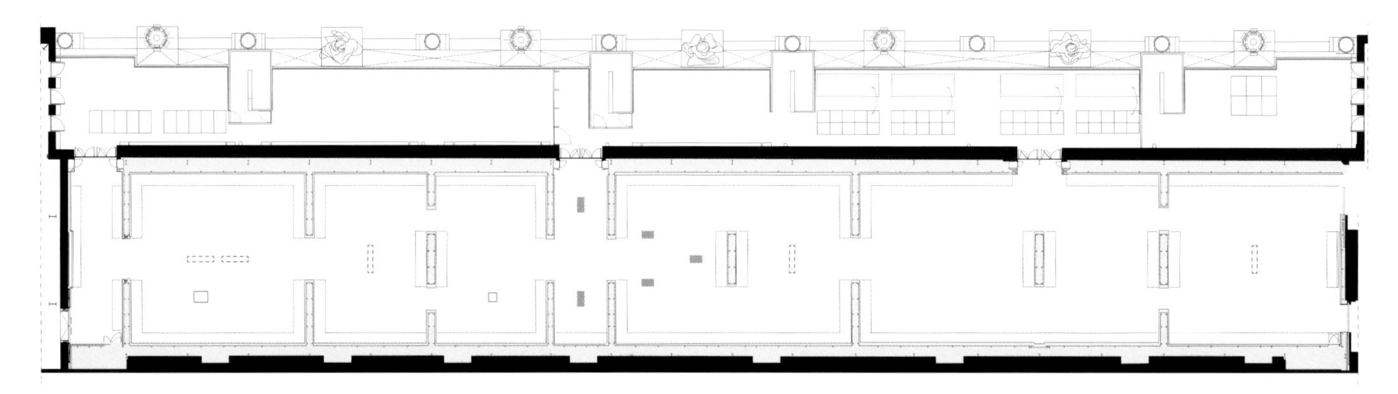

展柜布置图

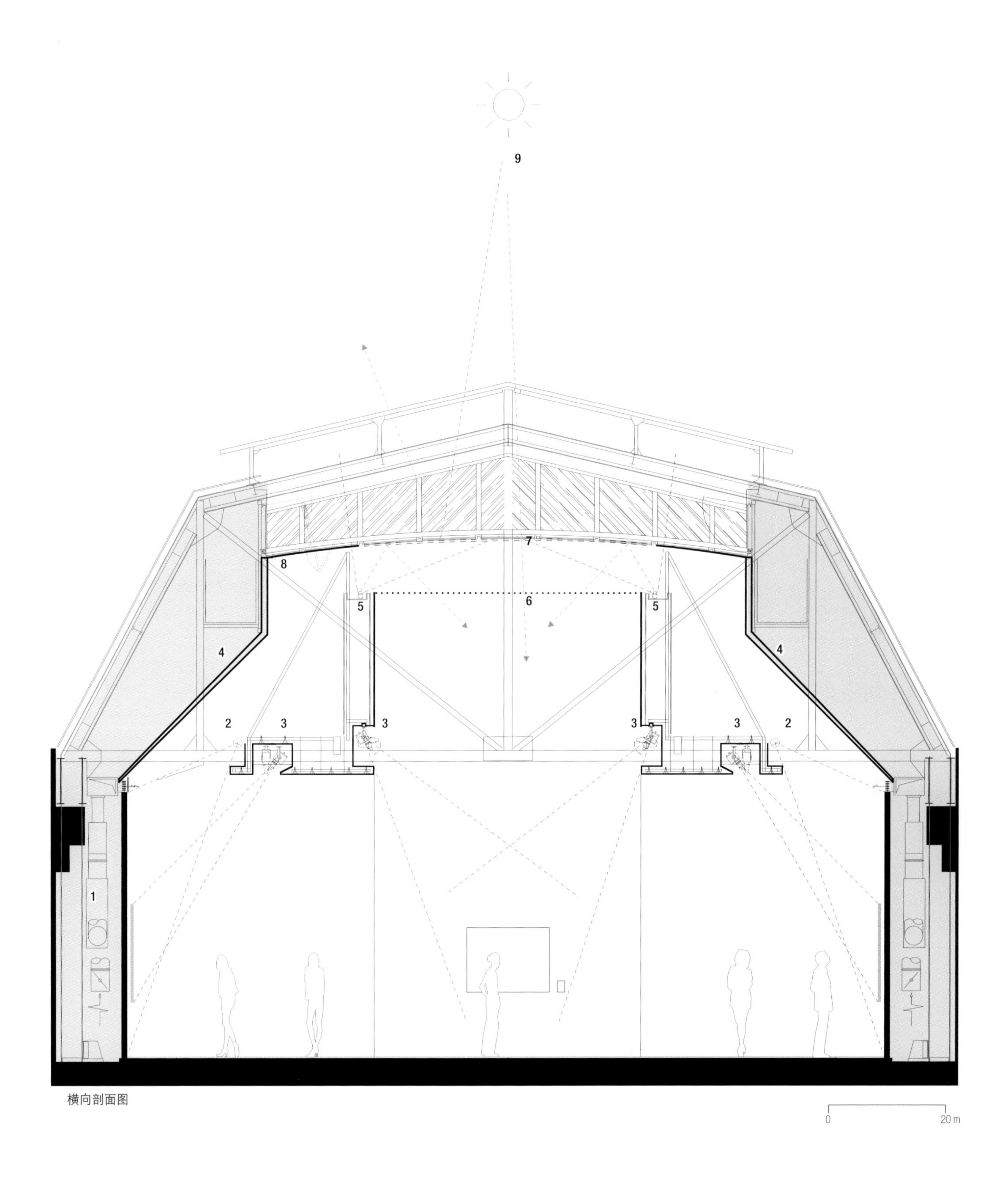

横向剖面图

1 服务设施管道
2 洗墙灯照明设施
3 聚光灯
4 黑色不透明薄膜
5 间接照明
6 金属网格天花板
7 双层薄膜：顶部反光薄膜或底部滤光薄膜
8 排烟设施
9 自然采光

Collection Mollard
Mollard Collection | Colección Mo

瑟索克博物馆

这个博物馆项目由威尔莫特建筑事务所与黎巴嫩建筑师雅克·阿布哈立德合作设计，瑟索克博物馆的翻新与扩建工程需要尽可能地谨慎，主要依靠地下的延伸部分进行。这座新摩尔式瑟索克博物馆建于1912年，当时是一座别墅，而如今则是以现代建筑为主的阿什拉菲耶区所剩无几的老房子之一。

该项目最初是在1998年构想的，但由于黎巴嫩的多年政治局势动荡该项目被搁置了下来，直到2008年才开始重新启动。博物馆项目的地下面积有8500平方米，其中7000平方米是新建的，这里包括一个临时的7米高的展览画廊。在可能的情况下尽量使用头顶自然采光方式，以白光LED系统作为补充。威尔莫特在项目中采用了沙黄色的埃及天然石材、玻璃和铝合金幕墙、可丽耐人造石和漂白橡木胶合镶板等。

建筑师将原来的花园规划与建筑入口进行了重组。一排被路径中断的树木恢复了整齐，迎面是一栋现代化的、拥有混凝土、金属与玻璃外表的建筑，建筑内部有一个图书馆、自助餐厅和一部汽车举升机。博物馆前的平坦空地围绕着六个中间带地灯的玻璃地板组成的路径。这个用石头铺成的花园中常年遍布、安放着各种雕塑。为了营造出新的展览空间以及存储空间、停车位、有160个座位的礼堂和多媒体图书馆等，在不破坏原有建筑的前提下，必须下挖超过20米。这是一个真正的挑战，深挖工程由当地的助理建筑师督导完成。自1961年以来，博物馆就一直在组织“艺术沙龙”，让艺术家们在此展示他们的作品的，同时也使这里成为卓越的原创艺术作品收藏之所。

项目地点 / 黎巴嫩贝鲁特
完工时间 / 2015年
建筑面积 / 8500平方米

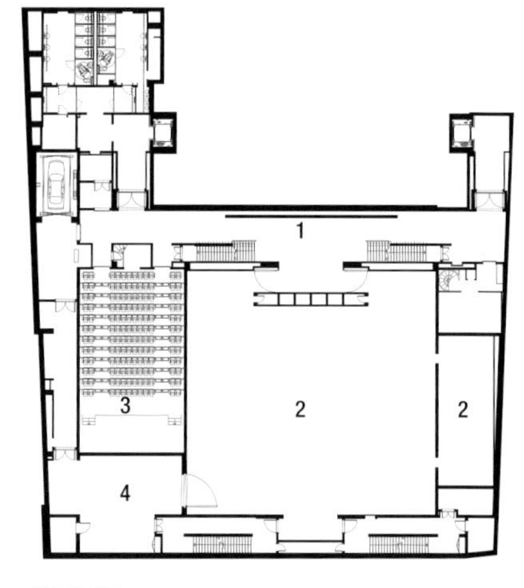

二层平面图

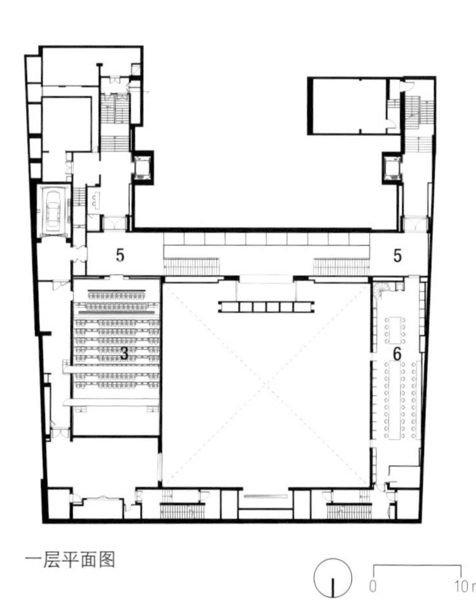

一层平面图

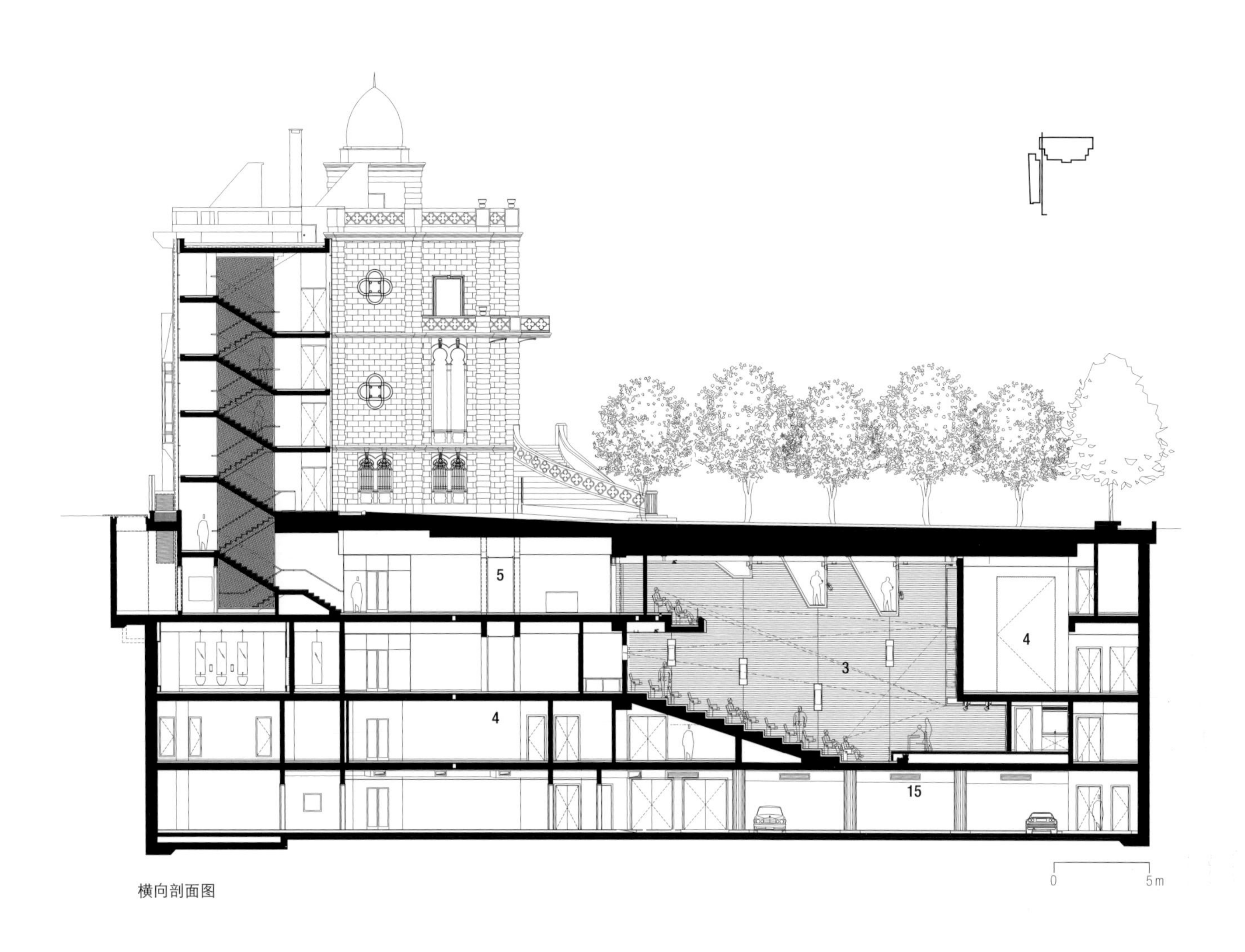

横向剖面图

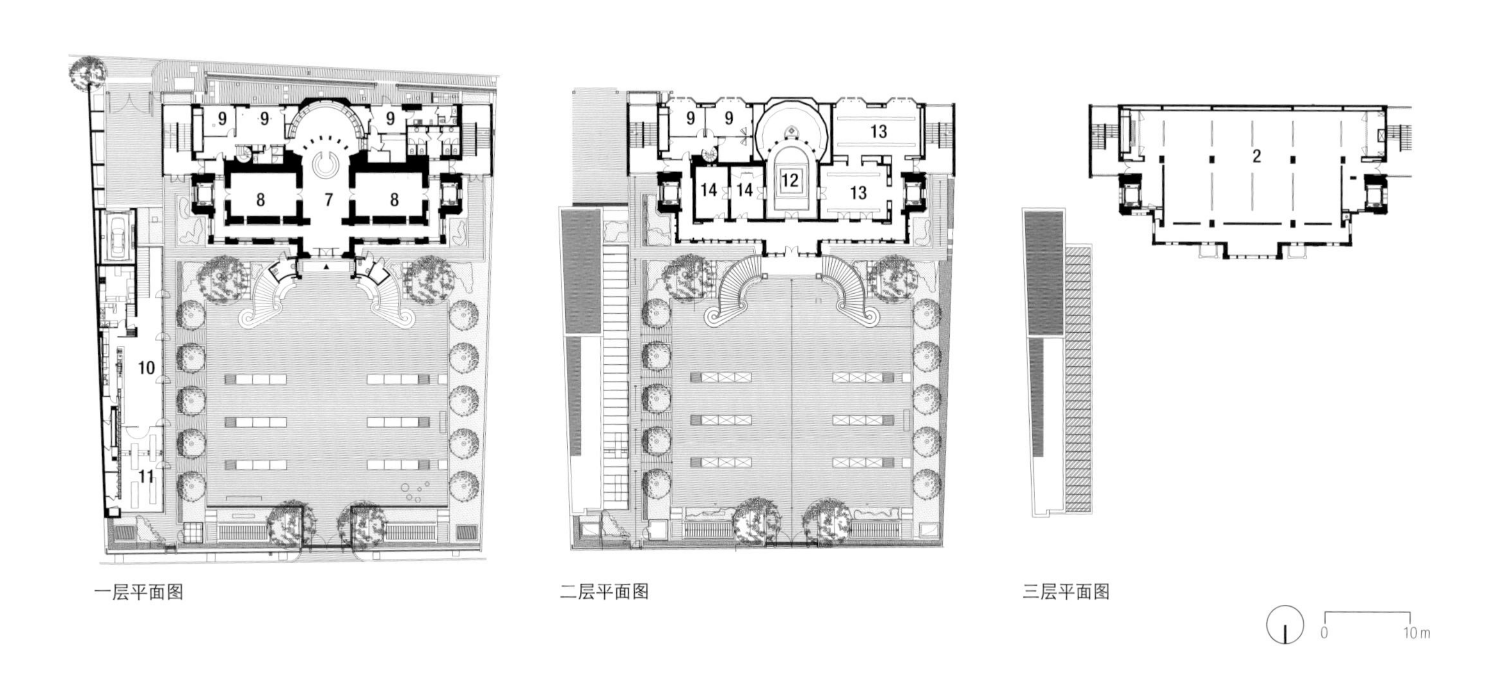

一层平面图

二层平面图

三层平面图

1 大厅
2 临时展览区
3 礼堂
4 仓储区
5 夹层楼
6 图书馆
7 门厅
8 画廊
9 办公室
10 餐厅
11 书店
12 阿拉伯式休息室
13 常设展览区
14 尼古拉斯·瑟索克办公室
15 停车场

荷兰国立博物馆扩建项目

荷兰国立博物馆建于1885年，由荷兰建筑师皮耶特・柯伊珀斯以新哥特式和文艺复兴风格建造的。2003年，荷兰政府实施了一个国立博物馆全面改建翻新计划，以便使它再次达到国际顶级博物馆的标准。总共12 000平方米的新展览空间由威尔莫特建筑事务所设计。西班牙克鲁兹&奥尔蒂斯建筑事务所则负责包括正门厅等在内的改建重组工作。

项目地点 / 荷兰阿姆斯特丹
完工时间 / 2013年
建筑面积 / 12 000平方米

威尔莫特的任务是恢复柯伊珀斯原来设计的内部结构和布局，并设计画廊空间，以使收藏品不会触及建筑物的墙壁。额外的设计考量是藏品的按年代顺序展览方案。展品的布置需要精心策划才能圆满地呈现艺术作品与文物并排摆放的鲜明对比。在设计中也不得不采取十分复杂的安全措施来保护这些无价之宝。

建筑师改造了这个建筑的四层楼：夹层楼放置中世纪的特别藏品与作品；一楼为18世纪和19世纪的收藏品；二楼是名作画廊和伦勃朗的《夜巡》，这也是荷兰国立博物馆馆藏的17世纪荷兰黄金时代无与伦比的艺术作品；三楼则陈列20世纪以后的艺术作品。新哥特式建筑内充斥着沉重的拱顶、丰富的装饰以及由柱子和拱门隔开的建筑结构，因此，将空间透明度与纯净度引入画廊的改造中所产生的效果是十分显著的。该团队特地为此增加了220个没有垂直框架的展柜以及防反光玻璃和精美的煤黑色金属材料。

威尔莫特为博物馆设计了一种独特的LED照明系统，由飞利浦照明开发，名为“Lightracks”，营造出画廊的主要色调——五阶灰度。阿姆斯特丹国立博物馆是迄今为止荷兰最重要的博物馆，也是世界上最伟大的博物馆之一，这都要感谢让-米歇尔・威尔莫特出色的工作。

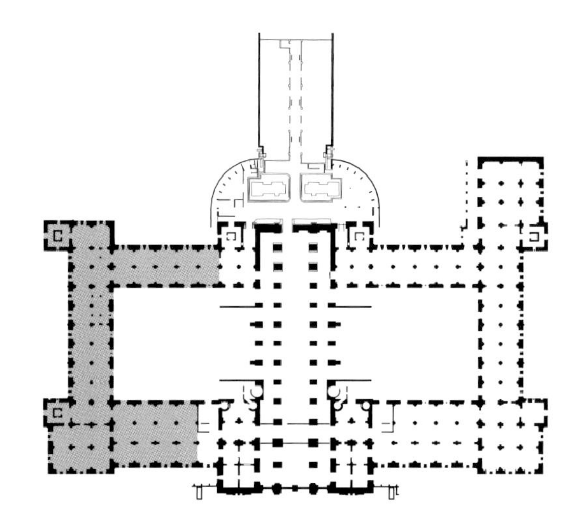

入口楼层特殊藏品分布位置图

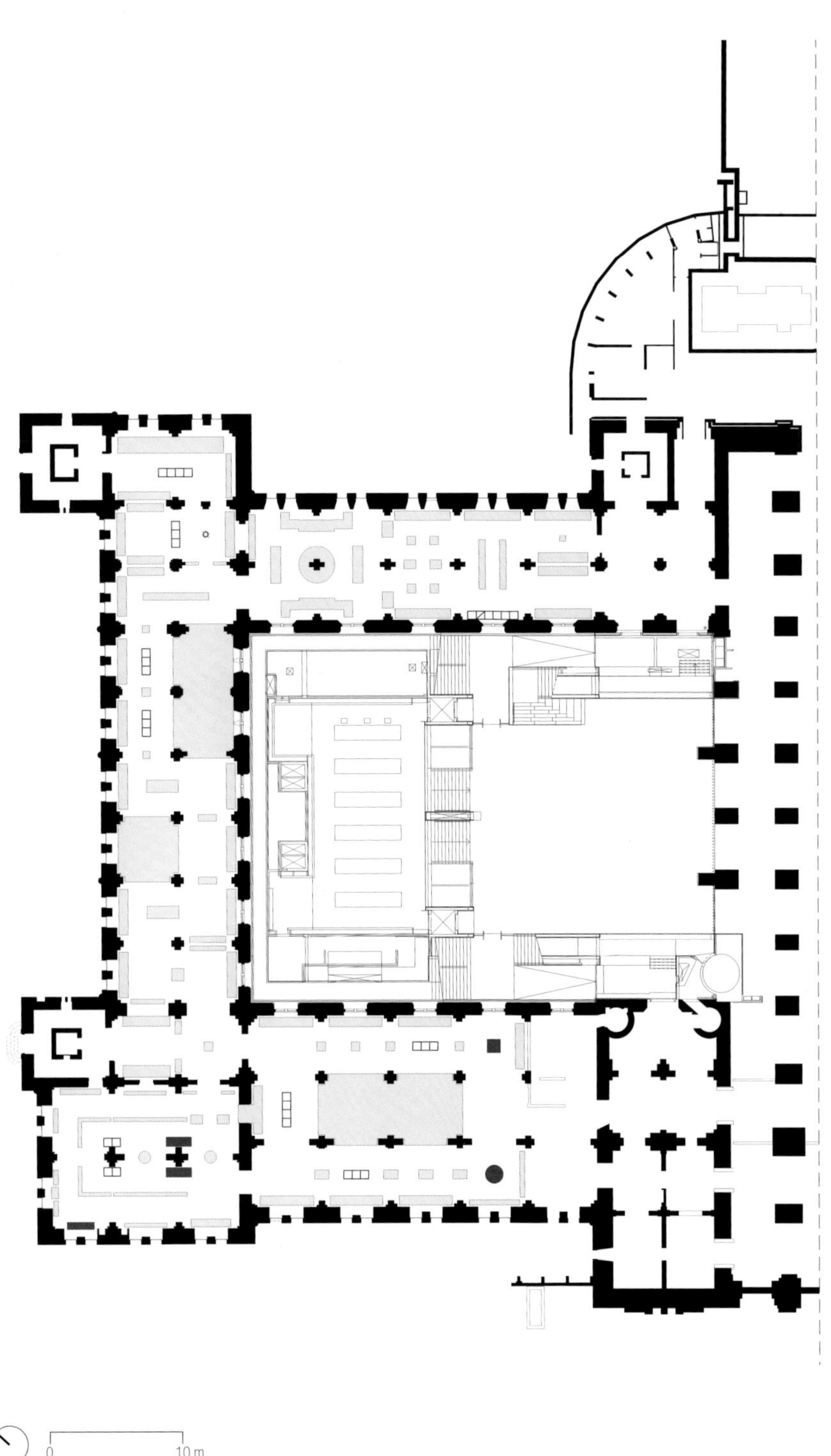

特别藏品展柜布置图

展台
独立式展柜
长椅

威尔莫特
建筑事务所简介

威尔莫特建筑事务所是一家国际化、跨文化、跨语种的建筑设计公司，是由城市建筑师、城市规划师、室内设计师让-米歇尔·威尔莫特于1975年创建的。

总部设在法国、在英国、意大利和韩国等地设有分支机构。威尔莫特建筑事务所目前拥有来自21个国家的，约250名建筑师、城市规划师、博物馆设计专家、室内设计师等，他们都是一些被求知探索、知识分享和创新创造的热情所驱动的人。其中六位设计师从建筑公司——威尔莫特设计工作室与工业公司分离出去之后，开始探索“环境设计”领域，并且在中国工业产品设计界也获得了良好的声誉。

40多年来，从达拉斯到巴黎，从里约热内卢到首尔，再到伦敦、达喀尔、威尼斯、巴黎和莫斯科，他们环游世界，寻找着新的机会之地。

让-米歇尔·威尔莫特和他的合作者们领导了27个不同国家的100多个项目，从最大的项目到最小的项目，从最壮观的项目到最平常的项目，从最初的草图到最后完工，他们都抱有同样的激情。

例如，他们近期完工的项目有法国俄罗斯东正教精神文化中心（法国，巴黎）、法国欧莱雅集团总部（法国，上塞纳省克利希镇）、联合利华集团总部（法国，上塞纳省吕埃马勒迈松镇）、日内瓦国际学校艺术中心（瑞士，日内瓦）、李考德路和福煦路的高档住宅楼（法国，凡尔赛）、韩国大田文化中心（韩国，大田市）、拥有36 200个座位的安联里维耶拉球场（法国，尼斯）、谷歌伦敦总部、德高集团（法）英国总部（英国，伦敦）、法拉利一级方程式运动跑车管理中心（意大利，马拉内罗）、圣保罗（巴西）展览中心展会设计项目和2015年欧洲运动会生态公园设计项目（阿塞拜疆，巴库），等等。

其他最近的项目任务还包括巴黎13区F站初创企业园区（世界上最大的初创企业园区）翻新改造项目、巴黎北站和奥斯特利兹车站翻新改造项目、法国梅斯会议中心项目、法国马赛遗传基因疾病研究所项目、圣彼得堡欧洲大学（俄罗斯）、利伯维尔新区（加蓬）项目，莫斯科住宅大楼项目（俄罗斯）、达喀尔办公大楼项目（塞内加尔）以及联合国西非迪亚姆尼亚迪奥总部（塞内加尔）等。

在这些既有创新性又有可持续性的项目中，设计总是要考虑到景观、照明、材料、装饰等方面，同时也要尊重建设地点当地的环境与历史情况。

威尔莫特工作室还和学生们与年轻建筑师们分享其知识、高规格的设计和价值观等，并设立PrixW设计大赛。这个建筑设计大赛于2005年开始举办，由威尔莫特基金会组织，专注于现代建筑与传统建筑的结合（建筑移植法），他们举办颁奖活动、发布设计作品，同时也在威尼斯建筑双年展期间用威尔莫特工作室自己的画廊展出获奖者作品。

自2010年以来，根据每年由英国《建筑设计》杂志发布的研究报告，威尔莫特建筑事务所已跻身世界建筑公司100强之中，并于2017年排名第73位。其设计的项目获得了许多荣誉与奖项，其中包括2014年度国际建筑奖（法国尼斯安联里维耶拉球场）、2015年度欧洲最佳博物馆奖（荷兰阿姆斯特丹国立博物馆）、2016年度巴黎首都大区大奖赛城市规划奖（F站初创企业园区）等。欧洲空间规划师理事会因威尔莫特建筑事务所的大鲁瓦西地城市战略规划也于2016年授予其年度欧洲城市规划奖，威尔莫特建筑事务所为该项目制定了用高速公路连接巴黎和鲁瓦西地区的新规划。

事务所地址

威尔莫特建筑事务所
68, rue du Faubourg Saint-Antoine
75012 Paris
FRANCE

威尔莫特实业公司
59, rue de Charonne
75011 Paris
FRANCE

威尔莫特建筑事务所（英国）
133, Oxford Gardens
London W10 6NE
UNITED KINGDOM

威尔莫特建筑事务所（意大利）
Cannarregio 3560
30121 Venice
ITALY

威尔莫特企业基金会
10, rue Sainte-Anastase
75003 Paris
FRANCE

公司网站
www.wilmotte.com

项目信息

建成项目

佳娜艺术中心

City/Country	Seoul/Republic of Korea (South Korea)
Completion date	1998
Client	Gana Art Center
Area	32,550 ft² (3,024 m²)
Architect	Wilmotte & Associés
Associate architect	Total Design
W&A's project manager	Fabrice Drain
W&A's team	Laurence Vandongeon, Alain Aubert, Astrid Courtois, Agathe de Kergos
W&A's program	Construction of exhibition spaces devoted to contemporary art; interior design, and museography
Building structure	Reinforced concrete
Building materials	Italian beige limestone, wood

李氏宅邸

City/Country	Seoul/Republic of Korea (South Korea)
Completion date	2005
Client	Private
Area	3,745 ft² (348 m²)
Architect	Wilmotte & Associés
Associate architect	Tetra Architects & Engineers
Landscape architect	Neveux-Rouyer
W&A's project manager	Eric Kittler
W&A's program	Construction and interior design of a private house
Building structure	Reinforced poured-in-place concrete
Building materials	Tamped concrete

板桥高档住宅

City/Country	Seongnam, Seoul/Republic of Korea (South Korea)
Completion date	2005
Client	IB Housing
Area	104,300 ft² (9,690 m²)
Architect	Wilmotte & Associés
Associate architect	Tetra Architects & Engineers
Landscape architect	Neveux-Rouyer
W&A's project managers	Nicolas Gilsoul/Jong-Ki Min/ Bénedicte Ollier
W&A's team	Jong-Hoon Shin, Eva Meinhardt, Colleen Caulliez
W&A's program	Construction of luxury apartment complex, seven buildings of four levels each (36 apartments in total); basement garages
Building structure	Reinforced concrete
Building materials	Italian white marble from Carrara, black Korean traditional brick for the ground, wood lattice, water

信号塔大厦

City/Country	La Défense, Paris/France
Competition date	2008
Client	EPAD
Area	1,237,850 ft² (115,000 m²)
Architect	Wilmotte & Associés
Landscape architect	Neveux-Rouyer
W&A's project manager	Ralf Levedag
W&A's team	Peter Hatzmann, Ossisane Caldicote, Lorenzo Gaetani, Miguel Pinto, Alexandra Fraczek, Barbara Jeanrond
W&A's program	Design of a 284-story ecological tower comprising offices, residential, hotel, tourist residence, shops and facilities

巴黎脑部与脊椎研究所

City/Country	Paris/France
Completion date	2010
Client	ADREC CHU Pitié Salpêtrière
Area	246,495 ft² (22,900 m²)
Architect	Wilmotte & Associés
W&A's project manager	Bettina Hoeptner
W&A's team	Fiona Copstick, Véronique Bouysse-Lacroix, Cristina Cordera, Alexandra Fraczek, Eugène Mathe
W&A's program	Construction of a building, including laboratories, patient bedrooms, human-imaging sector, and offices
Building structure	Reinforced concrete post-and-beam structure
Building materials	Gray-tinted glass curtain wall

蒙特卡洛观景塔大厦

City/Country	Monaco
Completion date	2012
Client	Michel Pastor Group
Area	107,640 ft² (10,000 m²)
Architect	Wilmotte & Associés
Associate architect	Alexis Blanchi
Landscape architect	Véronique Viale
W&A's project manager	Eric Kittler
W&A's team	Hector Montero, Tiphanie Priami Rasse, Rafik Samir Hamdi
W&A's program	Construction of a 19-story, 200-foot-tall (61-meter) high-rise tower
Building structure	Reinforced concrete
Building materials	Composite panels made of aluminum honeycomb with Moleanos natural stone surface and stainless steel profile façade)

拉德芳斯发展局总部大楼

City/Country	Nanterre/France
Completion date	2012
Client	BNP PARIBAS Promotion Immobilière
Area	161,460 ft² (15,000 m²)
Architect	Wilmotte & Associés
Landscape architect	Jean-Michel Rameau
W&A's project manager	Ralf Levedag
W&A's team	Cristina Cordera, Peter Hatzmann, Ossisane Caldicote
W&A's program	Construction of an office building
Building structure	Reinforced concrete, posts and bearing façade
Building materials	Aluminum front, double-glazed windows with solar control, wood decking
Awards	Pyramide d'argent 2012, Pyramide de Vermeil 2012

上诉法院律师培训学院

City/Country	Issy-les-Moulineaux/France
Completion date	2013, competition-winning project
Client	Bouygues Immobilier
Area	94,185 ft² (8,750 m²)
Architect	Wilmotte & Associés ,
Landscape architect	Neveux-Rouyer
W&A's project managers	Ralf Levedag, Charles Moliner, Olivier Chaubin
W&A's team	Jitendra Jain, Ludovic Thomas, Romain Gauthier, Christian Chehaiber, Woytek Sepiol
W&A's program	Construction of a public law school for 1,900 students
Building structure	Concrete and metallic structure
Building materials	Stone, wood, glass, steel, aluminum, hammered concrete

日内瓦国际学校艺术中心

City/Country	Geneva/Switzerland
Completion date	2014, restricted competition
Client	Foundation of the International School of Geneva
Area	50,915 ft² (4,730 m²)
Architect	Wilmotte & Associés
Associate architect	Brodbeck & Roulet (Marcel Hart, Daniel Zambon, and Mario Morgado)
Landscape architect	Gilbert Henchoz
W&A's project manager	Laurent Peyron
W&A's team	Edin Mujezinovic, Guillaume Renaud
W&A's program	Construction of an art center including a 400-seat multi-purpose auditorium, and a 200-seat theater
Building structure	Reinforced concrete
Building materials	Concrete, Linit U-profile glass by Lamberts

安联里维耶拉球场

City/Country	Nice/France
Completion date	2013, competition-winning project
Client	City of Nice
Area	35 acres (14 hectares)
Architect	Wilmotte & Associés
W&A's project managers	Ralf Levedag/Marco Punzi
W&A's team	Ossisane Caldicote, Louis Lafargue, Nicolas Lundström, Woytek Sepiol, Gabriela Richaud, Cristina Cordera, Peter Hatzmann, Véronique Bouysse-Lacroix, Jintendra Jain, Kuns Liu, Frank Avril, Julia Mille, Paul Bardon, Claire Rovella, Ilaria Ceccherini, Marine Gigout,

Agathe Bensa, Delphine Baldassini, Fred Bequelin, Lucie Sorrenti, Emmanuel Brelot, Jean-Luc Wagner, Marleen Homan, Céline Bourret, Edouard Debré, David Guyon, Anne-Claire Grassler, Xavier Turk, Cyril Lancelin, Guillaume Renaud, Romain Gauthier, Ali Nazemi, Alekos Santantonios, Gilles Sagot, Yong-Ki Hong, Seung Ki Kim, Alban Danguy des Déserts, Marc Dutoit, Vincent Leroy, Vanessa Adolphe

W&A's program Construction of a 35,000-seat stadium (UEFA approved) with multi-purpose facilities (sports and concerts), in a seismic zone

Building structure Concrete structure and triangulated timber/steel frame

Building materials Spruce, steel, concrete, ETFE

Awards International Architecture Award 2014

新国防部大楼象限角园区建筑

City/Country Paris/France

Completion date 2017, competition-winning project

Client SAS CorneOuest Promotion (Linkcity)

Area 992,440 ft² (92,200 m²)

Architect Wilmotte & Associés

Landscape architect Neveux-Rouyer

W&A's project managers Ralf Levedag/Jean Claude Antonio

W&A's team Frédéric Poinet, Hwan-Chul Kim, Xenia Spektor, Vassil Kaykov, Jean-Baptiste Heisse, Olivier Carminati, Lorraine Bertrand, Bettina Hoeptner, Ossisane Caldicote, Vincent Carbillet, Fiona Copstick, Francesca Gulizia, Raimundo Parraga, Romain Rivaux, Ludovic Thomas, Céline Bourret, Edouard Riou, Marie Lhuillier, Alexandre Maréchal

W&A's program Construction of four office buildings and a landscaped park for the Quadrans service campus on the site of the new French Ministry of Defence

Building structure Reinforced concrete post-and-beam structure

Building materials Concrete, stone, stainless steel, Corten steel, composite panels, thermo-coated aluminum, frame curtain wall, exterior glazing slide-in, terrace in exotic wood (ipe)

潘克拉斯6号方形大楼

City/Country London/United Kingdom

Completion date 2015, restricted competition by invitation—winning project

Client BNP Paribas Real Estate

Area 531,740 ft² (49,400 m²)

Architect Wilmotte & Associés

Executive architect Adamson Associates

Landscape architect Townshend Associates

W&A's project managers Malte Mager (competition)/ Pablo Lambrechts

W&A's team Benoit Sanson, Luca Vernocchi, Dhiren Patel, Matthew Brookes, Miguel Pinto, Hugo Loureiro, Carlos Manns, Romain Gauthier

W&A's program Construction of an office building and interior design of public areas

Building structure Reinforced concrete (basements and ground floor)/ steel construction with metal decks (other floors)

Building materials Terracotta, anodized aluminum, concrete, glass

大田文化中心

City/Country Daejeon/Republic of Korea (South Korea)

Completion date 2014, competition-winning project

Client City of Daejeon

Area 98,275 ft² (9,130 m²)

Architect Wilmotte & Associés

Associate architect Shinwha Architects & Engineering

W&A's project manager Michael Levy

W&A's team Yujong Choi, Yuhui Shi, Fiona Copstick, Mark Zerkaulen, Guillaume Renaud

W&A's program Construction of a cultural center including a 350-seat theatre, an exhibition hall, a music studio, offices, and parking lot

Building structure Reinforced concrete

Building materials Aluminum panels and glass

法拉利运动跑车管理中心

City/Country Maranello/Italy

Completion date 2015

Client Ferrari S.p.A.

Area 258,335 ft² (24,000 m²)

Architect Wilmotte & Associés

Associate architect Studio Quaranta

W&A's project manager Ralf Levedag

W&A's team Erica Bonomi, Lorenzo Gaetani, Nicolas Lundström, Giovanna Mancini, Matthew Brookes, Ossisane Caldicote, Warren Nicoud

W&A's program: Construction of the new Ferrari sporting management center devoted to the design and assembly of Ferrari Formula 1 racing cars
Building structure: Reinforced concrete
Building materials: Silk screened glass panels (ground floor façade), double-skin glass façade (upper floor façade)

俄罗斯东正教精神文化中心

City/Country: Paris/France
Completion date: 2016
Client: Russian Federation
Area: 50,105 ft² (4,655 m²)
Architect: Wilmotte & Associés
Landscape architect: Louis Benech
W&A's project manager: Ralf Levedag
W&A's team: Louis Lafargue, Therèse Leroy, Véronique Bouysse-Lacroix, Nicolas Lundström, Cyril Lancelin, Ossisane Caldicote, Warren Nicoud, Romain Rivaux, Romain Gauthier, Guillaume Renaud, Alekos Santantonios, Ali Nazemi, Jean-Baptiste Coursault, Gilles Sagot, Alexandre Thai
W&A's program: Construction of a cultural center, a cathedral, a spiritual center, and a primary school
Building structure: Reinforced concrete
Building materials: Massangis natural stone, glass, gold leaf

勃朗峰地区私人木屋

City/Country: France
Completion date: 2015
Client: Private
Area: 6,460 ft² (600 m²)
Architect: Wilmotte & Associés
Associate architect: Pascal Chatron-Michaud
Landscape architect: Neveux-Rouyer
W&A's project managers: Eric Kittler/Bertrand Leridon
W&A's team: Romain Nowak, Tiphanie Priami Rasse, Romain Gauthier, Alban Danguy des Déserts
W&A's program: Construction of a wood contemporary chalet
Building structure: Reinforced concrete, wood frame construction
Building materials: Textured stained oak, slate from Portugal, sheet metal

埃法日集团总部

City/Country: Vélizy-Villacoublay/France
Completion date: 2015
Client: Eiffage Immobilier
Area: 258,335 ft² (24,000 m²)
Architect: Wilmotte & Associés
Landscape architect: Neveux-Rouyer
W&A's project manager: Jean-Fançois Patte
W&A's team: Jean-Baptiste Heisse, Pierre Floch, Pierre Calame, Nancy Chidiac, Bénédicte Ollier, Romain Gauthier
W&A's program: Construction of the new Eiffage headquarters comprising offices, exhibition hall, library, company cafeteria, car park
Building structure: Concrete structure and steel frame
Building materials: Aluminum panel cladding, ventilated curtain wall, raw concrete
Awards: BIM d'Or 2014

珀夏尔大厦

City/Country: Dallas/USA
Completion date: 2017
Client: Harwood International
Area: 461,000 ft² (42,800 m²)
Architect: Wilmotte & Associés
Associate architect: Harwood Design Factory LLC
Landscape architect: Uchiyama Design Studio/Halff D+PC, Inc
W&A's project manager: Ralf Levedag
W&A's team: Louis Lafargue, Nicolas Lunsdström, Matthew Brookes, Ossisane Caldicote, Warren Nicoud
W&A's program: Construction of a 33-floor, 370-foot-tall (115-meter-tall) high-rise tower, including 158 luxury apartments
Building structure: Reinforced concrete
Building materials: Concrete, aluminum, glass, reconstituted stone

紫罗兰大厦

City/Country: Monaco
Completion date: 2020
Client: Michel Pastor Group
Area: 226,040 ft² (21,000 m²)
Architect: Wilmotte & Associés
Associate architect: Rainier Boisson
Landscape architect: Jean Mus & Compagnie
W&A's project manager: Ralf Levedag
W&A's team: Louis Lafargue, Claire Davisseau, Tudor Zamfirescu, Alban Danguy des Déserts, Ossisane Caldicote, Warren Nicoud

W&A's program Construction of a 22-floor, 425-foot-tall (130-meter-tall) residential high-rise tower, and six parking floors
Building structure Reinforced concrete
Building materials Stainless steel, stone, glass

翻新项目

瑞瑟夫别墅酒店

City/Country Ramatuelle/France
Completion date 2009
Client Michel Reybier
Area 89,340 ft^2 (8,300 m^2)
Architects Wilmotte & Associés
Landscape architect Neveux-Rouyer
W&A's project manager Anne-Claude Morand Dessart
W&A's team Jérôme Collet, Mio Shibuya, Chantal de Pelsmaeker
W&A's program Refurbishment of a 1970s building into a five-star hotel and spa, including seven rooms and 16 suites
Building structure Reinforced concrete
Building materials Coating concrete, dry stone, natural stone (flooring), wood decking
Awards Wallpaper Design Awards 2011, exclusive 'Distinction Palace' rating

让–皮埃尔·雷诺工作室

City/Country Barbizon/France
Completion date 2010
Client Jean Pierre Raynaud
Area 2,585 ft^2 (240 m^2)
Architects Wilmotte & Associés
Landscape architect Neveux-Rouyer
W&A's project manager Philippe Pingusson
W&A's program Restructuration and extension of an old Normandy-style house as an artist studio
Building structure Stonework covered in plastered filler imitating the freestone (an existing and partially conserved provisions at the back of the building), steel structure and glass curtain wall (new part of the building facing the wood)
Building materials Steel, aluminum, glass

文华东方酒店

City/Country Paris/France
Completion date 2011
Client Société Foncière Lyonnaise, Mandarin Oriental Hotel Group
Area 236,805 ft^2 (22,000 m^2)
Architects Charles Letrosne (1930s)/ Wilmotte & Associés
Interior architect SM Design/Jouin Manku
Landscape architect Neveux-Rouyer
W&A's project manager Christian Oudart
W&A's team Jean-Luc Perrin, Fabienne Dumant
W&A's program Restructuration of an office complex into a 5-star hotel including 99 rooms and 32 suites. First luxury hotel to receive HQE certification in France.
Building structure Reinforced concrete post-and-beam structure
Building materials Concrete, natural stone (façade on the street), natural stone and lime plaster (façade overlooking the garden)
Awards Exclusive 'Distinction Palace'rating

百德诗歌酒庄

City/Country Pauillac/France
Completion date 2015
Client Domaine Château Pédesclaux
Area 82,880 ft^2 (7,700 m^2)
Architects Wilmotte & Associés
Associate architect Atelier d'architecture BPM (Arnaud Boulain)
Landscape architect Neveux-Rouyer
W&A's project manager Laurent Brunier
W&A's team Elise Malisse, Angéline Geslain, Pierre-Yves Kuhn, Jean-Sébastien Pagnon, Romain Gauthier, Ali Nazemi
W&A's program Extension and refurbishment of Château Pédesclaux vineyards: creation of a cellar
Building structure White reinforced concrete (barrel warehouse in the basement) and bronze metallic structure (vat room on ground and first floor)
Building materials Black quartz concrete, bronze anodized aluminum strips (vat façade), glazed transparent front with large dimension glass panel, white concrete slabs, American walnut floor
Furniture Wilmotte & Industries and Victoria Wilmotte (tasting room)

德高集团英国总部

City/Country London/United Kingdom
Completion date 2016
Client DF Real Estate
Area 69,965 ft^2 (6,500m^2)
Architects Wilmotte & Associés

Landscape architect Townshend Associates
W&A's project managers Miguel Pinto/Pablo Lambrechts
W&A's team Jean-Luc Wagner
W&A's program Extension, refurbishment, interior design and re-cladding of JCDecaux offices
Building structure Existing concrete structure, extension and alterations in steel
Building materials Rendered concrete, aluminum, glass

F站初创企业园区

City/Country Paris/France
Completion date 2017
Client SDECN (Xavier Niel)
Area 365,975 ft^2 (34,000 m^2)
Architects Eugène Freyssinet (1927), historical listed building/ Wilmotte & Associés
Chief architect for historic monuments 2BDM Architectes
W&A's project manager Jean-François Patte
W&A's team Charlotte Bertrand, Jean Besson, Pierre Calame, Nancy Chidiac, Florian Giroguy, Théo Kirn, Marie Leyh, Carmen Villarta, Romain Queroy, Romain Gauthier, Alekos Santantonios, Ali Nazemi, Gilles Sagot, Jean-Baptiste Coursault, Cyril Lancelin, Guillaume Renaud, Ludovic Rubrecht, Alban Danguy des Déserts, Alexandre Thai
W&A's program Renovation to turn a monumental hall used for train-truck transfers of freight, built in pre-stressed concrete in 1927 by engineer Eugène Freyssinet, into a digital start-up hub
Building structure Pre-stressed concrete
Building materials Steel, wood, glass
Furniture Wilmotte & Industries

圣伯纳丁学院

City/Country Paris/France
Completion date 2008
Client Association diocésaine de Paris
Area 53,820 ft^2 (5,000 m^2)
Architect Wilmotte & Associés
Chief architect for historic monuments Hervé Baptiste
Landscape architect Neveux-Rouyer
W&A's project managers Emmanuel Brelot/XavierTurk
W&A's program Renovation of 13th- and 15th-century buildings, including the creation of two auditoriums and a library
Building structure Solid stone ancient structure, new roof with steel frame supporting the upper two floors; the entire load is supported by 300 sub-surface concrete micro-pilings
Building materials Saint-Maximin natural stone, steel joinery
Awards European Union Prize for Cultural Heritage, 2010 (Grand Prix)/Europa Nostra Awards (conservation)

城市规划项目

大莫斯科城市规划方案

City/Country Moscow/Russia
Completion date 2012, competition-winning project
Client City of Moscow
Area 258 mi^2 (668 km^2)
Architect Wilmotte & Associés
Architect planner Antoine Grumbach
Russian associate architect Serguei Tkachenko
W&A's project managers Charles Moliner/Anne Speicher/Jean-Pierre Franco Lucio
W&A's team Edouardo Iannicelli, Noémie Masson, Cyril Lancelin, Romain Gauthier, Guillaume Renaud, Gilles Sagot, Romain Queroy
W&A's program Master plan and mobility concept of Greater Moscow

让–饶勒斯大街项目

City/Country Nîmes/France
Completion date 2013, competition-winning project
Client City of Nîmes
Area 1 mile (1.6 km)
Architect and urban designer Wilmotte & Associés
Associate architect Carré d'Archi
Landscape architect Neveux-Rouyer
W&A's project managers Marie-Noëlle Passa/Anne Labroille/Marc Dutoit
W&A's team Ana Claudia Correa Diaz
W&A's program Urban design of Nîmes' largest avenue (1 mi-long [1.6 km], 200 ft wide [61 m])
Materials Sand-blasted concrete (sidewalk), concrete (parking area), concrete and stone slabs and paving stones made with natural stone from Croatia (central walkway), wooden duckboard (terrace)
Urban furniture Wilmotte & Industries
Awards Nominated project for the 2013 Urban Design Award (Moniteur Group), Victoire

du paysage 2016 'Grand Prix du Jury'

博物馆项目

卢浮宫原始艺术厅

City/Country Paris/France
Completion date 2000, competition-winning project
Client Etablissement public du Grand Louvre
Area 15,070 ft^2 (1,400 m^2)
Architect Hector-Martin Lefuel (1860s–1870s)
W&A's project manager Alain Desmarchelier
W&A's team Racha Kayali
W&A's program Museography
Showcase Wilmotte & Associés

雅克・希拉克总统博物馆

City/Country Sarran/France
Completion date 1998–2002 (Phase 1) & 2002–2006 (Phase 2)
Client Conseil général de la Corrèze
Area Phase 1: 16,145 ft^2 (1,500 m^2); Phase 2: 40,550 ft^2 (3,767 m^2)
Architect Wilmotte & Associés
Associate architects J. Parquet/D. Chassary (Atelier Centre); Nissou-Parquet
Landscape architect Michel Desvignes
W&A's project manager Emmanuel Brelot
W&A's team Xavier Turk, Alain Thibault, Yoo-jung Kim, Hélène Lecarpentier
W&A's program Refurbishment, interior design and subsequent extension of the museum to conserve gifts received by Jacques Chirac during his two terms as President of the French Republic
Building structure Reinforced concrete
Building materials Correzian pink granite, oak, coating, roofing natural slate

尤伦斯当代艺术中心

City/Country Beijing/China
Completion date 2007
Client Ullens Center for Contemporary Art
Area 80,730 ft^2 (7,500 m^2)
Architect Wilmotte & Associés
Associate architect MADA s.p.a.m.
W&A's project manager Emmanuel Brelot
W&A's team Xavier Turk, Moochol Shin, Jérôme Peuron, Myun Kim, Ho Keun Kwon
W&A's program Rehabilitation and refurbishment of a 1950s industrial building into an art gallery
Building structure Reinforced concrete
Building materials Fine epoxy resin (floor resin), painted wall

多哈伊斯兰艺术博物馆

City/Country Doha/Qatar
Completion date 2008
Client State of Qatar
Area 17,225 ft^2 (5,250 m^2)
Architect I. M. Pei Partnership Architects
Landscape architect Michel Desvignes
W&A's project manager Emmanuel Brelot
W&A's team Fabian Servagnat, Xavier Turk, Emilie Oliviero, Moochol Shin
W&A's program Museography and book shop/ gift shop furnishing
Showcases Wilmotte & Associés

奥赛博物馆翻新改造项目

City/Country Paris/France
Completion date 2011, competition-winning project
Client Etablissement public du Musée d'Orsay
Area 23,680 ft^2 (2,200 m^2)
Architect Victor Laloux (1898–1900)
W&A's project managers Emmanuel Brelot/Xavier Turk
W&A's program Museography and bookshop/ gift shop furnishing
Showcases Wilmotte & Associés

荷兰国立博物馆扩建项目

City/Country Amsterdam/Netherlands
Completion date 2013, competition-winning project
Client Rijksmuseum in partnership with the Ministry of Education, Culture and Science, and the Rijksgebouwendienst
Area 129,150 ft^2 (12,000 m^2)
Architects Pierre Cuypers (1885); Cruz y Ortiz (renovation)
W&A's project manager Marleen Homan
W&A's team Vanessa Adolphe, Marc Dutoit, Anne-Claire Grassler, Flore Lenoir, Domenico Lo Rito,Dji-Ming Luk, Emilie Oliviero, Bénédicte Ollier, Alekos Santantonios, Céline Seivert
W&A's program Museography
Furniture and showcases Wilmotte & Associés/Wilmotte & Industries
Awards Leaf Awards 2014, "international interior design" category

瑟索克博物馆

City/Country	Beirut/Lebanon
Completion date	1998–2015
Client	Nicolas Ibrahim Sursock Museum
Area	91,495 ft^2 (8,500 m^2)
Architects	Unknown/Wilmotte & Associés
Associate architect	JA Designs (Jacques Aboukhaled)
W&A's project managers	Alain Desmarchelier/Emmanuel Brelot
W&A's team	Ho Keun Kwon
W&A's program	Extension and refurbishment
Building structure	Reinforced concrete
Building materials	Egyptian natural stone, curtain wall (glass, aluminum), Corian, panel veneered in bleached oak

主要项目

进行中项目

法国

Antibes, Carrefour retail—refurbishment, 312,155 ft² (29,000 m²)

Jouy-en-Josas, HEC Campus—construction, 968,750 ft² (90,000 m²)

Ferney-Voltaire, Shopping Center—refurbishment and extension, 430,555 ft² (40,000 m²)

La Plagne Aime 2000, Hotel, residences, leisure center, retail—urban, architectural and landscape project, 581,250 ft² (54,000 m²)

Lyon, Les Terrasses de la Presqu'île—urban renewal, 6 acres (2.4 ha)

Marseilles, Giptis Research Institute—construction, 237,880 ft² (22,100 m²)

Metz, Congress Center—construction, 164,690 ft² (15,300 m²)

Neuilly-sur-Seine, American Hospital of Paris—construction of K building,168,995 ft² (15,700 m²)

Nice, Carrefour retail—extension, 172,225 ft² (16,000 m²)

Nice, IKEA store and residential—construction, 581,250 ft² (54,000 m²)

Orly, Coeur d'Orly—offices and retail, construction, 602,780 ft² (56,000 m²) (Askia offices, 193,750 ft²—18,000 m², 2015)

Paris, Lutetia hotel, 5-star—refurbishment, interior design and furniture design, 203,870 ft² (18,940 m²)

Paris-Saclay, Servier Laboratories—construction, 484,375 ft² (45,000 m²)

Paris-Roissy—modernization of A1 motorway between Paris and Roissy, 12 mi (20 km)

Poissy, PSG training center—construction, 699,655 ft² (65,000 m², interior) and 1,722,225 ft² (160,000 m², exterior)

Roquebrune Cap-Martin, Vista La Cigale hotel, 5-star—refurbishment, 113,020 ft² (10,500 m²)

其他国家

Congo (Republic), Loango, Slavery Memorial—construction, 199,130 ft² (18,500 m²)

Iran, Mashhad, International terminal of Mashhad Airport—construction, 398,265 ft² (37,000 m²)

Italy, Milan-Carrefour Paderno retail—construction, 430,555 ft² (40,000 m²)

Gabon, Libreville—masterplan and architecture, residential, offices, retail, coastline and marina, 268 acres (107 ha)

Luxembourg, Arcelor Mittal headquarters—construction, 570,487 ft² (53,000 m²)

Mauritius, Beau Champ-8 villas—construction,107,640 ft² (10,000 m²)

Monaco, Les Giroflées tower (residential)—construction, 226,040 ft² (21,000 m²)

Romania, Bucharest, Bioparc—masterplan, 143 acres (57 ha)

Russia, Moscow-Greater Moscow—development strategies, 258 mi² (668 km²)

Russia, Moscow, Red October (residential)—refurbishment, 441,320 ft² (41,000 m²)

Russia, Moscow, Red Side 2 towers (mixed use)—concept design, 1,114,065 ft² (103,500 m²)

Russia, St Petersburg, European University—refurbishment, 91,495 ft² (8,500 m²)

Saint-Tropez, Cheval Blanc-La Pinède hotel, 5-star (LVMH)—refurbishment and extension, 30,675 ft² (2,850 m²)

Senegal, Dakar, Mixed-use tower—construction, 340 ft high (104 m high), 222,815 ft² (20,700 m²)

Senegal, Dakar, Palace of Arts—refurbishment, 64,585 ft² (6,000 m²)

Senegal, Diamniadio, UN headquarters—construction, 430,555 ft² (40,000 m²)

USA, Dallas, Bleu Ciel tower (residential)—concept design, 461,000 ft² (42,800 m²); 5 towers (offices, residential, retail), 3,235,630 ft² (300,600 m²)

2015年前完成项目

法国

Paris, Gare du Nord—masterplan, 917,625 ft² (85,250 m²), 2016

Paris, Quadrans building—construction of offices on the site of the new Ministry of Defence, 992,440 ft² (92,200 m²) (Phase 1 completed in 2015, 492,985 ft² [45,800 m²]/Phase 2 completed in 2018, 499,445 ft² [46,400 m²])

Paris, Russian Orthodox Spiritual & Cultural Center—construction, 50,105 ft² (4,655 m²), 2016

Paris, Schlumberger headquarters—refurbishment, 43,055 ft² (4,000 m²), 2017

Paris, SMA headquarters and Okko hotel—construction, 419,790 ft² (39,000 m²), 2017

Paris, Station F Start-Up Campus—refurbishment, 365,975 ft² (34,000 m²), 2017

Rueil-Malmaison, Green Office, Unilever Headquarters—construction, 378,735 ft² (35,000 m²), 2015

Pauillac, Château Pédesclaux Winery—construction and extension, 82,880 ft² (7,700 m²), 2015

Versailles, Les Allées Foch—construction, 66,735 ft² (6,200 m²), 2015

Versailles, Les Allées Richaud—refurbishment and construction, 205,590 ft² (19,100 m²), 2015

Vélizy-Villacoublay, Eiffage headquarters—construction, 258,335 ft² (24,000 m²), 2015

其他国家

Azerbaijan, Baku, Baku Park for the European Games, 40 acres (16 ha), 2015

Brazil, Rio de Janeiro, Grand Mercure hotel, 4*—construction, 21,530 ft² (20,000 m²), 2015

Brazil, São Paulo, São Paulo Expo—construction, 1,065,625 ft² (99,000 m²), 2016

Italy, Maranello-Fiorano, Ferrari Sporting Management Center—construction, 258,335 ft² (24,000 m²), 2015

Lebanon, Beirut, Sursock Museum—refurbishment and interior design, 91,495 ft² (8,500 m²), 2015

South Korea, Daejeon, Cultural Center—construction, 98,275 ft² (9,130 m²), 2015

United Kingdom, King's Cross Central—Google London Headquarters—construction, 531,735 ft² (49,400 m²), 2015

United Kingdom, London, JCDecaux London Headquarters—refurbishment, 69,965 ft² (6,500 m²), 2016

South Korea, Seoul, Incheon International Airport, Terminal 1—interior design, 3,229,175 ft² (300,000 m²), 2000/Terminal 2—interior architecture concept and Art consultant, 4,090,286 ft² (380,000 m²), 2018

其他主要项目

法国

Bordeaux, Congress Center—refurbishment and extension, 117,325 ft² (10,900 m²), 2003

Clichy, L'Oréal headquarters—refurbishment and extension, 161,460 ft² (15,000 m²), 2014

Issy-les-Moulineaux, Barristers Training College—construction, 94,185 ft², (8,750 m²), 2013

Nanterre, EPADESA Corporate Headquarters—construction, 161,460 ft² (15,000 m²), 2012

Nice, Allianz Riviera sustainable stadium—construction, 581,250 ft² (54,000 m²), 36,200 seats, 2013

Nîmes, Allées Jean-Jaurès—urban design and furniture design, length 1 mile (1.6 km), width 200 ft (60 m), 2013

Paris, Avenue de France—urban design and street furniture, 2004

Paris, Collège de France—refurbishment and extension, 276,630 ft² (25,700 m²), 1998–2004

Paris, Collège des Bernardins—refurbishment and interior design, 53,820 ft² (5,000 m²), 2008

Paris, ICM (Brain and Spine Institute—construction, 246,495 ft² (22,900 m²), 2010

Paris, Maison de la Mutualité—Refurbishment, 139,930 ft² (13,000 m²), 2012

Paris, Mandarin-Oriental hotel, 5-star —refurbishment, 236,805 ft² (22,000 m²), 2011

Paris, Musée d'Orsay—refurbishment of 3 Impressionist galleries, 23,680 ft² (2,200 m²), 2011

Paris, Musée du Louvre:
—Grand Louvre, temporary exhibition galleries, shops, restaurants, 1988–1990
—Pavillon des Sessions, Department of Primitive Arts, interior design, 15,070 ft² (1,400 m²), 2000
—The Richelieu wing, interior design and specific furniture, 54,465 ft² (5,060 m²),1993
—The Rohan wing, interior design, 10,765 ft² (1,000 m²), 1999

Paris, Tramway des Maréchaux Sud, RATP T3—urban furniture, 2006

Ramatuelle, La Réserve hôtel, 5-star hotel-spa—refurbishment and interior design, 89,340 ft² (8,300 m²), 2009
Rueil-Malmaison, Schneider Electric headquarters—construction, 365,975 ft² (34,000m²), 2008

Saint-Estèphe, Château Cos d'Estournel Winery—refurbishment and extension, 57,048 ft² (5,300 m²), 2008

Sophia-Antipolis, SophiaTech Campus—construction, 197,520 ft² (18,350 m²), 2012

其他国家

China, Beijing, UCCA—refurbishment of a plant into a Contemporary Art Center, 80,730 ft² (7,500 m²), 2007

Italy, Forlì, San Domenico Museum—refurbishment and interior design, 53,820 ft² (5,000 m²), 2006

Monaco, Monte Carlo View Tower, residential—construction, 107,640 ft² (10,000 m²), 2012

South Korea, Seoul, Gana Art Center—construction, 32,550 ft² (3,024 m²), 1998

South Korea, Seongnam, Pangyo—construction of 36 flats, 104,302 ft² (9,690 m²), 2005

The Netherlands, Amsterdam, The New Rijksmuseum—interior design and museography, 12,915 ft² (12,000 m²), 2013

Portugal, Lisbon, Chiado Museum—refurbishment and museography, 36,600 ft² (3,400 m²), 1994

Qatar, Doha, Museum of Islamic Art—interior design and museography 17,225 ft² (5,250 m²)

Switzerland, Geneva, International School—construction of Arts Center, 50,915 ft² (4,730 m²), 2014

荣誉奖项

2017 GESTE D'OR, *Architecture, Materials and Research Grand Prix*, awarded to Station F, the world's biggest start-up campus, Paris, France.

2017 PYRAMIDES D'OR, *First Achievement Award*, awarded to the Quai de la Borde residence project in Ris-Orangis.

2017 CNCC TROPHY AT SIEC 2017, *design of a retail park award*, awarded to L'Avenue 83 project in La Valette-du-Var.

2017 PYRAMIDES D'ARGENT ILE-DE-FRANCE, Low Carbon building Award, awarded for the Épicéa residence project in Issy-les-Moulineaux.

2017 PYRAMIDES D'ARGENT ILE-DE-FRANCE, First Achievement Award, awarded for the Quai de la Borde residence project in Ris-Orangis.

2016 5E ÉDITION DES VICTOIRES DU PAYSAGE, *Grand Prix du Jury* awarded to the Allées Jean-Jaurès project, Nîmes, France.

2016 Jean-Michel Wilmotte was rewarded by a BFM Awards for his whole career.

2016 GRANDS PRIX DE LA RÉGION CAPITALE, *Urban design prize* awarded to the rehabilitation of the Halle Freyssinet into Station F, the world's biggest start-up campus, Paris, France.

2016 GRAND PRIX EUROPÉEN DE L'URBANISME awarded by the European Council of Town Planners for the territory of the Grand Roissy. Wilmotte & Associés is associated to this project through the requalification of the A1 motorway linking Paris to Roissy.

2016 JANUS DU COMMERCE, awarded by the Institut Français du Design to L'Avenue 83 retail park, La Valette-du-Var, France.

2016 iF DESIGN AWARD (PRODUCT), awarded by the International Forum Design GmbH, GRAFA lighting system manufactured by Artemide.

2016 PIERRES D'OR, Programme category, awarded to the refurbishment of the Halle Freyssinet into Station F, the world's biggest start-ups campus, Paris, France.

2015 TROPHÉES DU LOGEMENT ET DES TERRITOIRES, Sustainable programme of the year prize awarded to Quai de la Borde residential building, Ris-Orangis, France.

2015 PRIX NATIONAL DE LA CONSTRUCTION BOIS (National Prize of the Wooden Construction), 2nd prize awarded to the Allianz Riviera stadium, Nice, France.

2015 GRAND PRIX DU RAYONNEMENT FRANÇAIS (French Influence Award), cultural influence prize awarded to Jean-Michel Wilmotte, Paris, France.

2015 BEST EUROPEAN MUSEUM, awarded by the European Museum Forum to the Rijksmuseum, Amsterdam, Netherlands.

2015 PYRAMIDES D'ARGENT ILE-DE-FRANCE, Prix Spécial du Jury, Refurbishment and transformation of the former royal hospital of Versailles into residential, retail, cultural spaces and offices, Versailles, France.

2015 PYRAMIDES D'ARGENT MIDI-PYRÉNÉES, Prix d'Immobilier d'Entreprise awarded to Safran office buildings, France, Toulouse.

2015 14th edition of the PRIX AMO (ARCHITECTURE & clients) LIEU DE TRAVAIL ARCHITECTURE ENVIRONNEMENT, Special prize awarded for the refurbishment and modernization of the L'Oréal group headquarters, Clichy, France.

2014 BIM D'OR (BUILDING INFORMATION MODELING AWARD awarded to the Campus Pierre Berger, Eiffage headquarters, Velizy-Villacoublay, France.

2015 LES ÉTOILES DE L'OBSERVEUR DU DESIGN awarded for the design of URBA, street lamp manufactured by Thorn.

2014 LE GESTE D'OR has distinguished the rehabilitation of the Halle Freyssinet into Station F, the world's biggest start-up campus, Paris, France.

2014 LEAF AWARDS, International Interior Design category, awarded by Leaf International to The New Rijksmuseum, Amsterdam, The Netherlands.

2014 INTERNATIONAL ARCHITECTURE AWARD, awarded by The Chicago Athenaeum & The European Center of Architecture Art Design and Urban Studies to Allianz Riviera stadium, Nice, France.

2013 LES ÉTOILES DE L'OBSERVEUR DU DESIGN, "Made in France" prize awarded to the kitchen appliances LA CORNUE W manufactured by La Cornue.

2013 JANUS DE LA CITE, Éco Design mention, awarded by the Institut Français du Design to the Paris dustbin Bagatelle manufactured by Seri.

2012 GOOD DESIGN AWARD awarded by the Athenaeum-Museum of Architecture and Design to the ATOLLO washbasin manufactured by Rapsel.

2012 PYRAMIDES D'ARGENT ILE-DE-FRANCE, Prix de l'immobilier d'entreprise awarded to Via Verde building, EPADESA headquarters, Nanterre, France.

2011 GOOD DESIGN KOREA awarded by the Korea Institute of Design Promotion to Samsung group for the Samsung Raemian village project, Yongin, South Korea.

2011 MAN OF THE YEAR, awarded to Jean-Michel Wilmotte by the wine magazine *La Revue du Vin de France* for the Château Cos d'Estournel winery, Saint-Estèphe, France.

2011 WALLPAPER DESIGN AWARDS, BEST NEW HOTEL prize awarded to La Réserve 5-star hotel-spa, Ramatuelle, France.

2010 PRIX EUROPA NOSTRA awarded for the best example of European renovation and heritage conservation for the refurbishment of the Collège des Bernardins, Paris, France.

2010 TROPHÉE D'OR OF FESTIVAL FIMBACTE awarded to André Malraux Mediathèque, Béziers, France.

2009 STRATEGIES, GRAND PRIX DU DESIGN 2009 Commercial architecture Prize awarded to the Docks 76 retail park, Rouen, France.

2007 PYRAMIDES D'ARGENT ILE-DE-FRANCE: Grand Prix Regional awarded for the residential programme Résidences du Mail at Bois d'Arcy, France (Real Estate Developer: Promogim) & PRIX DE L'ESTHETIQUE IMMOBILIERE awarded for the residential programme *Côté Coeur* à Issy-les-Moulineaux, France (Real Estate Developer: Sefri-Cime).

2006 ARTURBAIN.FR award during the Robert Auzelle seminar with the support of Dominique Perben, Ministry of Transports, Equipment, Tourism and Sea, awarded for the Place de la Libération, Dijon, France.

2006 IF DESIGN AWARD awarded by the International Forum Design for the lighting systems *iWay* and *iRoad* manufactured by iGuzzini.

2004 IF DESIGN AWARD awarded by the International Forum Design for the lighting system *PhaosXeno* LED manufactured by Zumtobel.

1997 PRIX CITEC awarded to the exhibited collection at the CITEXPO 97 show, Montpellier, France.

1996 EUROPEAN PARKING AWARD awarded in Budapest by the European Parking Association to the Parking des Célestins (Lyon Parc Auto), Interior design: J.-M Wilmotte, M. Targe, D. Buren, Lyon, France.

1995 LA LAMPE DE TRENTE ANS during the 30th edition of the International Lighting exhibition at Porte de Versailles, Paris, France.

1994 GRAND PRIX NATIONAL DE LA CREATION INDUSTRIELLE awarded by the French Ministry of Culture.

1992 PRIX D'AMENAGEMENT ET D'URBANISME, public space category awarded by the Publisher *Le Moniteur des Villes* for the urban renewal of the city of Agde with the creation of specific furniture, France.

1991 OSCAR DU DESIGN awarded by the *Nouvel Economiste* for the Espace Technal, Toulouse, France.

1990 PRIX EUROPA NOSTRA awarded to the refurbishment of the Grenier à Sel as the best example of European renovation and heritage conservation, Avignon, France.

1990 MEDAILLE D'ARCHITECTURE INTERIEURE awarded by the Académie d'architecture, Paris, France.

1989 PRIX S.A.D (Salon des Artistes Décorateurs), France.

1989 Elected DESIGNER OF THE YEAR during the International Furniture Exhibition in Paris, France.

1988 LE LIN D'OR award, Milan, Italy.

1985 LAMPE D'OR awarded during the International Lighting Exhibition in Paris for the Washington lamp.

荣誉与头衔

Member of the *Cercle Pierres d'Or* since 2016

Member of the *Académie des beaux-arts*, Architecture department since 2015

Member of the board of Directors of the BBCA (low-carbon building) association since 2015

Administrator of the *Arts Décoratifs*, Paris, since 2008

Dean of the University of Architecture of Hongik, South Korea 2006–08

Chevalier de l'Ordre National de la Légion d'Honneur 1999

Grand Cross of *Ordem do infant dom Henrique*, awarded by the President of Portugal Mário Soares 1994

Chevalier de l'Ordre National du Mérite 1994

Commandeur des Arts et des Lettres 1992

Chevalier des Palmes Académiques 1986

出版物精选

让-米歇尔·威尔莫特的文章

'Dessin d'acier, dessein sublimé' in *Guide de la réhabilitation avec l'acier à l'usage des architectes et des ingénieurs*, Saint-Aubin, CTICM, 2010, pp. 6–7.

Dictionnaire amoureux de l'Architecture, Paris, Editions Plon, 2016, p. 787.

'L'acier est bien vivant' in *Guide de la réhabilitation des enveloppes et des planchers*, Eyrolles, Paris, 2014, p. 7.

'Question de bon sens' in *Architecture Intérieure des Villes. Interior, urban, design*, Le Moniteur des Villes, Paris, 1999, pp. 14–15.

'Regards émotions…' [sur Jean Lurçat], in *Jean Lurçat* (1892–1966): *Au seul bruit du soleil*, Cinisello Balsamo (Italy), Silvana Editoriale, 2016, p. 15.

'Retrouver Venise' in *Venezia, la scomparsa*, Paris, Editions Xavier Barral, 2017, p. 5.

Wilmotte, J.-M., Jodidio, P. 'Voir l'espace autrement' in *Connaissances des Arts*, n°480, February 1992, pp. 70–79.

作品集

Alvarez, J. *Wilmotte*. Paris, Editions du Regard, 2016.

Bony, A., Jodidio, P. Wilmotte. *Design: 1975–2015*, Munich, Prestel, 2019 [forthcoming book].

Direction de la Culture et de l'Animation de la ville de Saint-Quentin, Jean-Michel Wilmotte, architect/designer, Saint-Quentin, 2002, Catalogue of the 'Jean-Michel Wilmotte, architecte/designer' exhibition.

Grisoni, J., Loubeyre J.-B. *Wilmotte, L'instinct architecte*, Le Passage, Paris, 2005.

Institut culturel français, *Wilmotte, Elogio dell'evidenza*, AFAA, 1996. Catalogue of the Italy traveling exhibition (Florence, Rome, Forli, Bari), June 1996–August 1998.

Lamarre, F., Tournaire P. (photos), *Jean-Michel Wilmotte, Architectures à l'œuvre*, Le Moniteur, Paris, 2008.

McDowell, D. *Jean-Michel Wilmotte, Architecture-Écritures*, Aubanel, Geneva, 2009.

Meier, R. (foreword), Barré, F. (preface) *Wilmotte. Réalisations et projets*, Le Moniteur, Paris, 1993 (reprinted 1995).

Pradel, J.-L. *Wilmotte*. Electa Moniteur, Milan, Paris, 1989.

Rambert, F. Jean-Michel Wilmotte, Editions du Regard, Paris, 1996.

Wilmotte & Associés, *Projets Récents Futurs*, London, PUSH, 2009.

Wilmotte, J.-M., Virilio, P. (foreword) *Architecture Intérieure des Villes*, Interior, urban, design, Le Moniteur des Villes, Paris, 1999.

与威尔莫特建筑事务所项目相关的作品

Amar, M., Bianchi, F. Maison de la Mutualité, Paris, 2012.

Banque de Luxembourg. *Architects–Arquitectonica*, Blue Print Extra 12, London, 1994.

Besacier, H. Désveaux, D., Meyronin, B., Rambert, F., Collomb, G. (foreword), LPA (François Gindre), Groupe 45 (Georges Verney Caron), Ceci n'est pas un parc [Lyon's car parks], Libel, 2010.

Bony, A. *Meubles et décors des années 80*, Editions du Regard, Paris, 2010.

Colbert Management & Conseil, *Les Demeures Anahita by Wilmotte*, Colbert Management & Conseil, Paris, 2016.

Crespolini, C., Godon, P. [Rouen] Métrobus, éditions La Vie du Rail, Paris, 1995.

Dagen, P., Béret, C., Lévy, B.-H. Paix. Clara Halter – Jean-Michel Wilmotte. Editions Cercle d'Art, Paris, 2005.

Desmoulins, C. (dir.), *Le Collège des Bernardins, Histoire d'une reconversion,* Editions Alternatives, Paris,2009.

Dutertre, P. 'Place du Grand-Jardin, gare routière à Chelles' in *Paysages urbains, une France intime*, le Moniteur, Paris, 2007, pp. 90–91.

Engel, P. *Guide de la réhabilitation des enveloppes et des planchers*, Eyrolles, Paris, 2014.

Engel, P. *Guide de la réhabilitation avec l'acier à l'usage des architectes et des ingénieurs*, CTICM, Saint-Aubin, 2010.

Engel, P., Berlanda, T., Bruno, A., Mazzolani, F. 'Rénover et réinvestir l'existant' in Construire en acier, N° 3+4, 2010, pp. 4–17.

Engel, P., Koolhas, R. (foreword) *Manuel de la réhabilitation avec l'acier à l'usage des architectes et des ingénieurs*, Presses de l'EPFL, Lausanne, 2017.

Eric Mazoyer, E., Du Govic, N. *Paroles d'architectes*, Bouygues Immobilier, 2002.

Henriques da Silva, R. (dir.), *Obraçom: Museu do Chiado. Historias vistas e contadas*, Instituto Português de Museus, Lisbon, 1995.

Hugonot, M.-C. 'Jean-Michel Wilmotte: spécialiste de la diversité' in *Habiter la montagne. Chalets et maisons d'architectes*, Editions Glénat, Grenoble, 2016. pp. 58–67.
'Jean-Michel Wilmotte, Hotel de Nell and La Réserve Ramatuelle' in *The design hotels book 2016*, Design Hotels AG, Berlin, 2016, pp. 230–234/252–255.
Jodidio, P. 'Chalet Greystone. Courchevel, France' in House with a view, residential mountain architecture / Vue d'en haut, residences de montagne, Victoria (Australia), Images Publishing, 2008, pp. 246–249.
Jodidio, P. 'Château de Méry-sur-Oise' in *Connaissance des Arts*, special issue, 2001.
Jodidio, P. 'Jean-Michel Wilmotte. Frejus St Raphaël Community heater. Tour du Ruou house' in *Architecture in France*. Taschen, Cologne, 2006, pp. 184–191.
Jodidio, P. 'Jean-Michel Wilmotte: Ullens Center for contemporary Art (UCCA)' in *Architecture Now 6*, Cologne, Taschen, 2009, pp. 546–551.
Jodidio, P. 'Jean-Michel Wilmotte: Bordeaux Lac Convention Center' in *Architecture Now 3*, Taschen, Cologne, 2004, pp. 534–539.
Jodidio, P. 'Jean-Michel Wilmotte: Contemporary Art Center, Hamon Donation' in *Architecture Now 2*, Taschen, Cologne, 2002, pp. 566–571.
Jodidio, P., Lammerhuber, L. (photos), Museum of Islamic Art, Doha, Qatar, Munich, Prestel, 2008.
Jodidio, P. 'Lyon: le Musée des beaux-Arts' in *Connaissance des Arts*, special issue, 1994.
Jodidio, P. 'Wilmotte. Muséographie' in *Connaissance des Arts*, special issue, 1994.
'L'acier de fond en comble [Collège des Bernardins]' in *Construire en acier*, N° 3+4, 2010, pp. 30–35.
Lamarre, F. 'L'œuvre au noir [Restructuring of Lyon's customs building]' in *Europe, acier, architecture*, n°8, April 2008, pp. 40–43.
Leloup, M., *De la Halle Freyssinet à la Station F*, Paris, Editions Alternatives, 2017.
Maubant, J.-L. *Lyon: la ville, l'art et la voiture*, Villeubanne, Art/Edition, 1995.
Mc Dowell, D. 'Un coin d'ombre et de douceur sur la plage' [about Ramatuelle's villa by Jean-Michel Wilmotte], in *Jean Mus: jardins méditerranéens contemporains*, Editions Ulmer, Paris, 2016, pp. 164–171.
Mollard, C., Le Bon, L. *L'art de concevoir et gérer un musée*. Antony: Editions du Moniteur, 2016.
Müller, D. 'Multi-faceted: Museum of Islamic Arts in Doha/Qt' in *Professional Lighting Design*, n° 68, Sept./Oct. 2009, pp. 33–39.
Peyrel, B., Quinton, M. *Altarea Cogedim: Regards de créateurs* [Wilmotte: Cœur d'Orly, Promenade de Flandre (Roncq), Site Safran (Toulouse), l'Avenue 83 (Toulon-La Valette)], Editions de la Martinière, Paris, 2014.
Wilmotte, J.-M., Pradel J.-L. (foreword), *Mémoire d'empire. Trésors du Musée national du Palais de Tapei. Mise en scène par Wilmotte*, Arles, Actes Sud/AFAA, 1999.
Zabalbeascoa, A. 'Jean-Michel Wilmotte's architecture studio, Paris, 1991' in *The architect's office*, Barcelona, Editorial Gustavo Gili, 1996. pp. 172–175.

威尔莫特基金会出版物

Catalogue of the Prix W

Prix W 2018. Fort de Villiers, Paris, Fondation Wilmotte, 2018.
Prix W 2016. Pondorly, Paris, Fondation Wilmotte, 2016.
Prix W 2014. Tower of London, Paris, Fondation Wilmotte, 2014.
Prix W 2012. Industrial site in Venice, Paris, Fondation Wilmotte, 2012.
Prix W 2010. Depot of the archives of the Bibliothèque nationale de France in Versailles, Fondation Wilmotte, Paris, 2011.
Prix W 2009. Water tower in Latina, Fondation Wilmotte, Paris, 2009.
Prix W 2007. Minoterie Lepoivre, éditions Jean-Michel Place, Paris, 2007.
Prix W 2006. Château Barrière, éditions Jean-Michel Place, Paris, 2006.

展览

Bruno, A., Wilmotte J.-M. (editorial), Jodidio, P. Fare, Andrea Bruno: Disfare, Rifare Architettura. Da Rivoli a Bagrati, Paris, Fondation Wilmotte, 2014. Exhibition catalogue, Venice (Italy), from September 18, 2014 to January 30, 2015.
Calandre, P. Isola Nova, Paris, Fondation Wilmotte, 2014. Exhibition catalogue of photographs by Philippe Calandre. Venice (Italy), from December 18, 2013 to May 20, 2014. London (United Kingdom), from September 13, 2014 to November 15, 2014.
Delangle, F. (photos), Wilmotte, J.-M., Ponti, S. (texts), Venezia, la scomparsa, Paris, Editions Xavier Barral, 2017, p. 136 Exhibition catalogue of photographies by Frédéric Delangle. Venice (Italy), from May 11 to November 26, 2017
Galassi, G. (photos), Ginapri, L.(text) Elogio della Luce: visioni del Razionalismo italiano, Fondation Wilmotte, 2015. Exhibition catalogue of photographs by Gianni Galassi. Venice (Italy), from November 7, 2015 to February 28, 2016.

威尔莫特作品展览精选

'Etat d'Esprit'
Institut Français d'Architecture/Paris (France), from November 19 to December 13, 1986.

'Jean-Michel Wilmotte'
Yamagiwa Inspiration Gallery/Tokyo (Japan), June 1–10, 1992. *Exhibition of Wilmotte's furniture, urban furniture and architecture*

'Novator 93'
Saint-Martin-ès-Aires Chapel/Troyes (France), from June 17 to August 31, 1993. *Exhibition of lighting items*

'Jean-Michel Wilmotte'
Rouen (France), May 1994. *Photographs of Wilmotte's urban furniture and architecture. Exhibition organized by Rouen's SIVOM for the occasion of the presentation of 22 Rouen's Metrobus stations design*

Jean-Michel Wilmotte at 'Art du Jardin' Fair
Saint-Cloud (France), from May 30 to June 3, 1996. *Projects of benches and street lights designed for Hess Form + Licht*

'Jean-Michel Wilmotte'
Seoul Art Center/Seoul (South Korea), September 3–28, 1996. *Exhibition of selected works*

'Jean-Michel Wilmotte'
Modern Art Museum/Taipei (Taiwan), Autumn 1996. *Exhibition of selected works*

'Jean-Michel Wilmotte'
Tokyo Bunkamura/Tokyo (Japan), Spring 1997. *Exhibition of selected works*

'Jean-Michel Wilmotte au Salon des Maires et des collectivités locales'
Parc des Expositions/Paris (France), November 19–21, 1996. *Exhibition featuring benches and street lights designed by Wilmotte for Hess Form + Licht in the auditorium and the main walkway of the Paris Parc des Expositions at Porte de Versailles*

'Architecture intérieure des villes'
68, rue du Faubourg Saint-Antoine, Paris (France), from November 18, 1997 to 1998. *Presentation of urban design projects by Jean-Michel Wilmotte and staging of the new urban furniture collection designed for Hess Form + Licht company*

'Esterni – Interni'
Doge's Palace/Venice (Italy), from September 25 to October 28, 1998. *Exhibition of selected works*

'Elogio dell' evidenza'
French cultural institutes/Florence, Rome, Forli, Bari (Italy), from June 18, 1996 to 30 August 30, 1998. *Exhibition of selected works. Itinerary exhibition supported by the AFAA (Ministry of Foreign Affairs) and the French Cultural Institute presented in several cities of Italy*

'Architecture intérieure des villes'
Gana Art Gallery/Seoul (South Korea), from September 4 to October 4, 1999. *Exhibition of a selection of urban planning projects and urban furniture*

'Fonte de ville'
L'Arc Scène Nationale/Le Creusot (France), from October 8 to December 19, 1999. *Exhibition featuring benches and street lights designed by Wilmotte for Hess Form + Licht, and presentation of photographs: interplay of light and shade*

'Jean-Michel Wilmotte – Selected projects'
Hôtel de Région/Le Puy-en-Velay (France), 2002. *Exhibition of selected works in architecture, museography, urban planning and design*

'Jean-Michel Wilmotte – Selected projects'
French embassy/Athens (Greece), 2004. *Exhibition of selected works in architecture, museography, urban planning, and design*

'Nouvelle ligne de mobilier urbain pour Hess'
Bordeaux (France), 2004. *Exhibition of a new line of urban furniture during the Week of Architecture*

'Jean-Michel Wilmotte, architect/designer'
Galerie Saint-Jacques, Saint-Quentin (France), from April 19 to May 15, 2002. *Exhibition of selected works in architecture, museography, urban planning and design. Jean-Michel Wilmotte received the title of 'Citizen of Honor of Saint-Quentin'*

'Wilmotte UK Launch Event'
London (United Kingdom), 2009. *Exhibition of selected works in architecture, museography, urban planning, and design*

'Utopies et Innovations'
Saint-Louis (France), 2010. *Presentation of the Wilmotte Foundation and selected works of Wilmotte & Associés*

Designer's days, Opening of the 'Galerie d'actualités Wilmotte'
Galerie d'actualités Wilmotte/Paris (France), from May 31 to June 4, 2012. *Presentation of the new collection of high-end kitchen equipment designed for La Cornue W; presentation of the Nice Stadium project and Eclatec lighting installation on the Passerelle des Arts*

'Design for U.S.' Paris Design Week 2012
Galerie d'actualités Wilmotte/Paris (France), September 10–16, 2012. *Presentation of the new furniture collection designed for American company Holly Hunt*

'Aplat by wilmotte'
Galerie d'actualités Wilmotte/Paris (France), January 2014. *Presentation of a concept of high-performance and environmentally friendly architectural paints, named 'Aplat by Wilmotte,' in partnership with Jefco, manufacturer of professional paintings*

'Architecture Passions, 40 ans de créations Wilmotte & Associés'
Espace Richaud/Versailles (France), September to November 27, 2016. *Presentation of Wilmotte & Associés design practice and a number of selected projects*

Exhibitions organized by the Wilmotte Foundation

'Prix W 2012'
Wilmotte Foundation/Venice (Italy), from September 29 to November 25, 2012. *Exhibition of the winning projects of the Prix W 2012 award & presentation of a selection of Wilmotte & Associés' projects*

'Otto mani e un occhio: Nasser Bouzid, Jean-Gabriel Coignet, Côme Mosta-Heirt, François Perrodin, Marin Kasimir'
Wilmotte Foundation/Venice (Italy), from May 31 to November 24, 2013. *Exhibition dedicated to four sculptors and one photographer*

'Isola Nova. Philippe Calandre'
Wilmotte Foundation/Venice (Italy), from December 18, 2013 to May 20, 2014. Wilmotte UK/London (United Kingdom), from September 13, 2014 to November 15, 2014. *Presentation of a series of new islands in Venice inhabited by large industrial structures mixed with fragments of traditional Venetian architecture*

'TraFumetto e gioco: disegniveneziani. Matteo Alemanno, Marco Maggi e Francesco Nepitello'
Wilmotte Foundation/Venice (Italy), from April 5 to May 20, 2014. *Presentation of original drawings from the author of strip cartoons Matteo Alemanno, and also Venetian games*

'Prix W 2014: Design a cultural and events center for the Tower of London'
Wilmotte Foundation/Venice (Italy), from May to September 2014. *Exhibition of the winning projects of the Prix W 2014 award*

'Andrea Bruno: Fare, Disfare, Rifare Architettura, da Rivoli a Bagrati'
Wilmotte Foundation/Venice (Italy), from September 18, 2014 to January 30, 2015. *Retrospective of Italian architect from Turin Andrea Bruno's work*

'Relectures. Alessandra Chemollo'
Wilmotte Foundation/Venice (Italy), from May 7 to October 25, 2015. Wilmotte UK/London (United Kingdom), from June 8 to November 30, 2016. *Exhibition of some of Jean-Michel Wilmotte's projects in Museum design, interpreted by Alessandra Chemollo, photographer*

'Elogio della luce [Praise of light]. Gianni Galassi'
Wilmotte Foundation/Venice (Italy), from November 7, 2015 to February 28, 2016. *Gianni Galassi captures the purity of the Italian rationalist architecture through magnificent and almost abstract photographs*

'Prix W 2016: Pondorly, design of a new arc de Triomphe for Paris'
Wilmotte Foundation/Venice (Italy), from May 6 to September 15, 2016. *Exhibition of the winning projects of the Prix W 2016 award*

'Russian Orthodox Cathedral of Paris'
Wilmotte Foundation/Venice (Italy), from February 18 to March 2, 2017. *Presentation of the Russian Orthodox Cathedral of Paris designed by Wilmotte & Associés*

'Venezia, la scomparsa. Frédéric Delangle'
Wilmotte Foundation/Venice (Italy), from May 11 to November 26, 2017. *Commissioned by Jean-Michel Wilmotte, presentation of French photographer Frédéric Delangle's work dedicated to a new vision of Venice in 36 original prints*

Exhibitions commissioned by Jean-Michel Wilmotte

'Elogio dell' ombra. Bruno Romeda'
68, rue du Faubourg Saint-Antoine/Paris (France), from January 28 to February 17, 1993. *Presentation of Italian sculptor Bruno Romeda's work*

'The ruins of Detroit. Yves Marchand & Romain Meffre'
Wilmotte UK/London (United Kingdom), from February 27 to April 27, 2012. *Presentation of the photographs of Detroit, Michigan by French photographers Yves Marchand and Romain Meffre*

'Art in progress. Leonora Hamill'
Wilmotte UK/London (United Kingdom), from February 15 to March 28, 2013. *For the first time in the United Kingdom, presentation of an exhibition dedicated to British photographer Leonora Hamill's work, winner of the 'Prix HSBC pour la Photographie (2012)'*

'Wild Surfaces. Edith Marie Pasquier'
Wilmotte UK/London (United Kingdom), from June 19 to July 26, 2013. *For the first time in the United Kingdom, presentation of British photographer Edith Marie Pasquier's work*

'Le temple du Soleil. Patrizia Mussa'
Wilmotte UK/London (United Kingdom), from May 22 to September 20, 2015. *Presentation of the photographs of the Grande Motte made by Italian artist Patrizia Mussa*

'Ubiquity: 2. Miguel Chevalier'
Wilmotte UK/London (United Kingdom), from April 13 to June 15, 2018. *Interactive visual installation by Miguel Chevalier*

图片版权信息

书中所有位置平面图、设计平面图和其他技术图纸均由威尔莫特建筑事务所提供，由波林・亨利、安东尼奥・马里诺和多纳塔・布金斯凯特重新绘制。

新建项目

李氏宅邸 © Pascal Tournaire 18–20; © Ki-Hwan Lee 21

板桥高档住宅 © Ki-Hwan Lee 24–25, 26; © Pascal Tournaire 27

信号塔大厦 © Roomservice 3d – Niels Kretschmann 30–31, 32, 33

巴黎脑部与脊椎研究所 © Didier Boy de la Tour 36–37, 38, 39

蒙特卡洛观景塔大厦 © Milène Servelle 42 (top), 43; © Pascal Pronnier 42 (bottom)

拉德芳斯发展局总部大楼 © Didier Boy de la Tour 46–47

上诉法院律师培训学院 © Didier Boy de la Tour 52; © Nicolas Fussler 50, 51, 53

日内瓦国际学校艺术中心 © Adrien Barakat 56–57

安联里维耶拉球场 © Milène Servelle 60–65

新国防部大楼象限角园区建筑 © Antoine Huot 68–69; © Nicolas Fussler 70–71

潘克拉斯6号方形大楼 © Steve de Vriendt 74, 75; © Edmund Sumner Partnership 75, 76

大田文化中心 © Ki-Hwan Lee 80–81, 82, 83

法拉利运动跑车管理中心 © Milène Servelle 86–87, 88, 89

俄罗斯东正教精神文化中心 © Alessandra Chemollo 92–93, 94, 96–97; © Mario Fourmy 95

勃朗峰地区私人木屋 © Milène Servelle 100–103

埃法日集团总部 © Nicolas Fussler 106–109

珀夏尔大厦 © Joseph Haubert 112; © Wilmotte & Associés 113

紫罗兰大厦 © Wilmotte & Associés 116, 117

翻新项目

瑞瑟夫别墅酒店 © Jacques Denarnaud 122–123, 127; © Grégoire Gardette 124, 126 (top); © Manuel Zublena 125, 126 (bottom)

让–皮埃尔・雷诺工作室 © Françoise Huguier/VU' 130–131; © Philippe Chancel 132–133

文华东方酒店 © Didier Boy de la Tour 136–137, 139; © Guillaume Maucuit Lecomte 138

百德诗歌酒庄 © Rodolphe Escher 142–143, 144; © Anaka 145

德高集团英国总部 © Edmund Sumner Partnership 148–151

F站初创企业园区 © Patrick Tourneboeuf 154–161

圣伯纳丁学院 © Pascal Tournaire 164, 165, 167; Géraldine Bruneel 166 (top), 169; Pierre-Olivier Deschamps/VU' 166 (bottom); Guillaume Maucuit Lecomte 168

城市规划项目

大莫斯科城市规划方案 © Wilmotte & Associés, Antoine Grumbach and Sergueï Tkachenko 173, 174, 176, 177; © Wilmotte & Associés, Antoine Grumbach, Sergueï Tkachenko and Atelier Villes et Paysages 175

让–饶勒斯大街项目 © Didier Boy de la Tour 180–181, 183; © Dominique Marck 182

博物馆项目

卢浮宫原始艺术厅 © Didier Boy de la Tour 188–189, 191; © Alessandra Chemollo 190

佳娜艺术中心 © Pascal Tournaire 194; © Jung Chea Park 195

雅克・希拉克总统博物馆 © Didier Boy de la Tour 200 (top); © Pascal Tournaire 198–199, 200 (middle and bottom), 201

尤伦斯当代艺术中心 © André Morin/Yan Pei-Ming 204–205; © Pascal Tournaire 206–207

多哈伊斯兰艺术博物馆 © Lois Lammerhuber 210, 212, 213; © Zed Photography 211

奥赛博物馆翻新改造项目 © Alessandra Chemollo 216–221

瑟索克博物馆 © Emmanuel Brelot 228–9; © Géraldine Bruneel 230, 231

荷兰国立博物馆扩建项目 © Alessandra Chemollo 230–231, 233; © Julien Lanoo 232, 234–235

项目索引

出版商尽最大努力确认了本书中所有图片的版权来源，同时欢迎版权所有者来信更正版权错误或遗漏。书中的信息和图片均由让-米歇尔·威尔莫特建筑事务所准备并提供。虽然在出版过程中已尽量确保内容的准确性，但对于书中出现的错误与遗漏等问题，出版商概不负责。